REAL BODIES

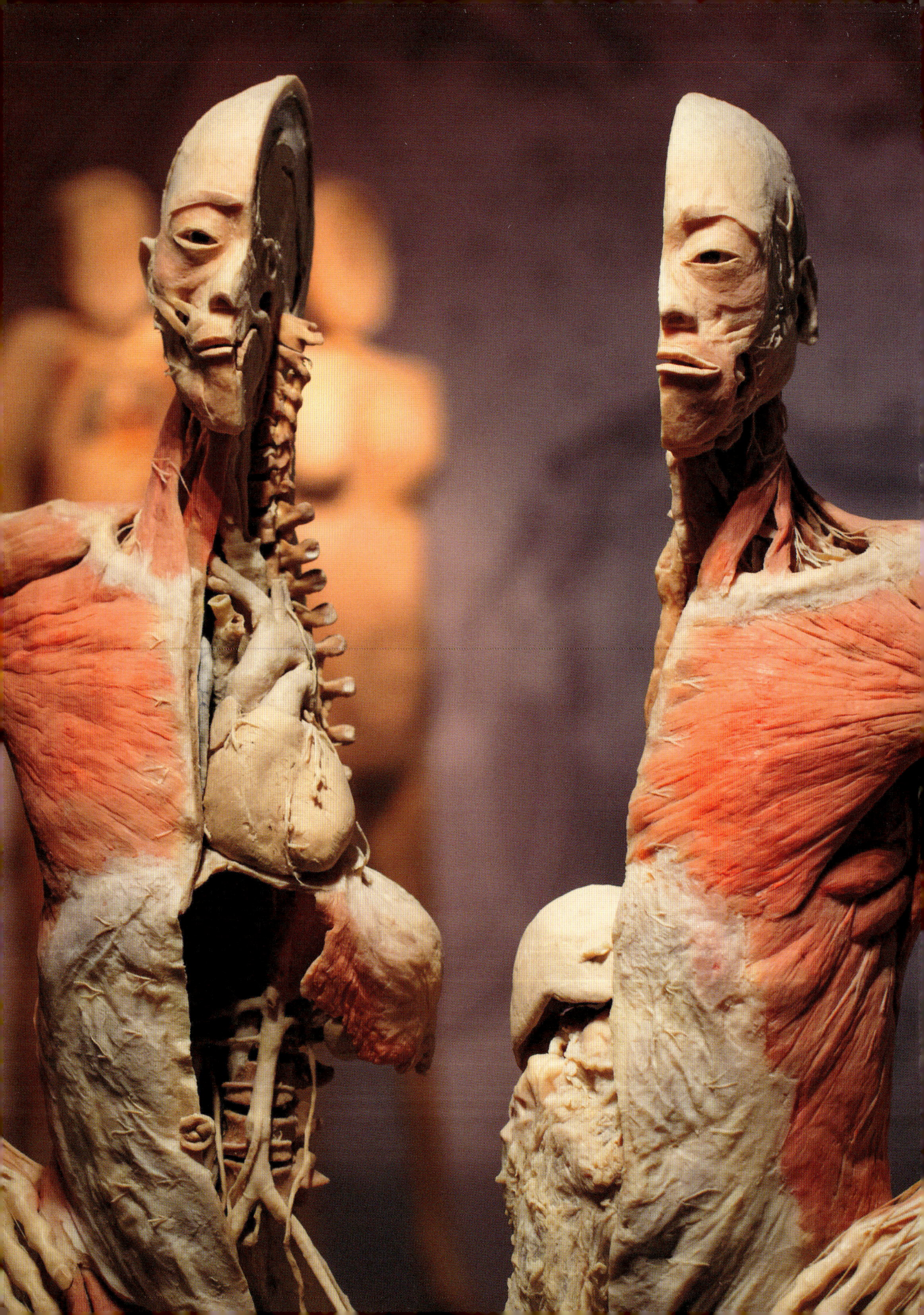

ABOUT THE EXHIBITION

WHAT ARE WE MADE OF?
WHERE DO WE COME FROM?
WHY ARE WE HERE?

These questions have persisted since modern humans drew their first breath. At different times in our lives these questions take on different forms–popularity in school, love in our twenties, life goals in our thirties, career in middle age, our legacy as we grow old–yet they all come from the same place: a desire to know who we are and to understand what is going on inside of us. This booklet is your chance to take a deeper look.

REAL BODIES uses real human specimens to explore the complex inner workings of the human form, the only thing we all share. Following a system-by-system approach, this Exhibition will give you greater insight into what's going on inside of you. Along the way, you'll also explore how elements such as breathing, hunger, and the rhythm of the heart have deep cultural and emotional significance that reaches back as far as the dawn of humanity.

We are not just organisms. We are beings, more complex than the systems within us, more powerful than the will of our anscestors who ensured we made it to today, more magical than the mysteries of birth and death.

So we begin our exploration into flesh, bone, blood, tissue and cell, but also beyond all of these. We are rooted in our past, in our beliefs, and in an unrelenting drive to know more.

Let's see what we're all about.

ANATOMIST'S STUDY

REAL BODIES uses an advanced form of preservation that allows you to study the inner workings of the human form in great detail. However, preservation of human remains has been in practice for thousands of years. The ancient Egyptians are the best known culture to practice the dehydration of bodies through mummification. A 70 day process, mummification removes most internal organs except the heart and slowly dries out the skin. Many other cultures also practiced mummification including the Aztecs, Mayans, and the native peoples of what are now Chile and Peru, who first used the practice nearly 8,000 years ago. This type of preservation was primarily used for religious purposes, but also meant that the priests who performed the elaborate rituals to create mummies were well versed in human anatomy.

Another form of preservation comes from a naturally occurring phenomenon in which body tissue deposited in airless, humid environments transforms into a natural type of wax, essentially sealing the body and preventing decay. Some of these bodies have been recovered over the centuries (as early as the 1200s) and displayed more as objects of curiosity rather than objects of study.

As interest in the inner working of the human body increased, anatomists and natural scientists looked for ways to preserve specimens for longer term study. This included creating "wet specimens," which are preserved in a heavy liquid or chemical bath to prevent decay, and the modern day practice of embalming in which all body fluids are replaced with formaldehyde or its concentrated version of formalin. The goal in these forms of preservation was often to retain typical and unique specimens for in depth analysis and learning. The Anatomist's Study you are about to enter celebrates the spirit of endless exploration into our inner workings practiced by so many individuals over thousands of years–each with the quest to better understand the human body. Such inquiry, which often stemmed from religious practices and included comparative anatomy (the study of animals), grew into highly focused investigation of the human body. The knowledge accumulated over centuries and millennia by the desire to know what makes us tick has led, and continues to lead us to a clearer understanding of our inner workings and how to heal us.

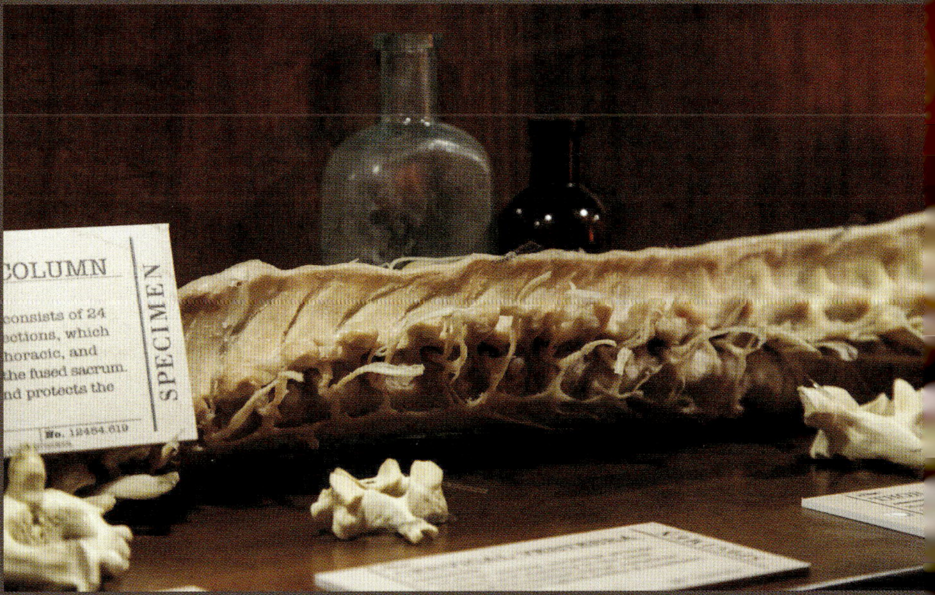

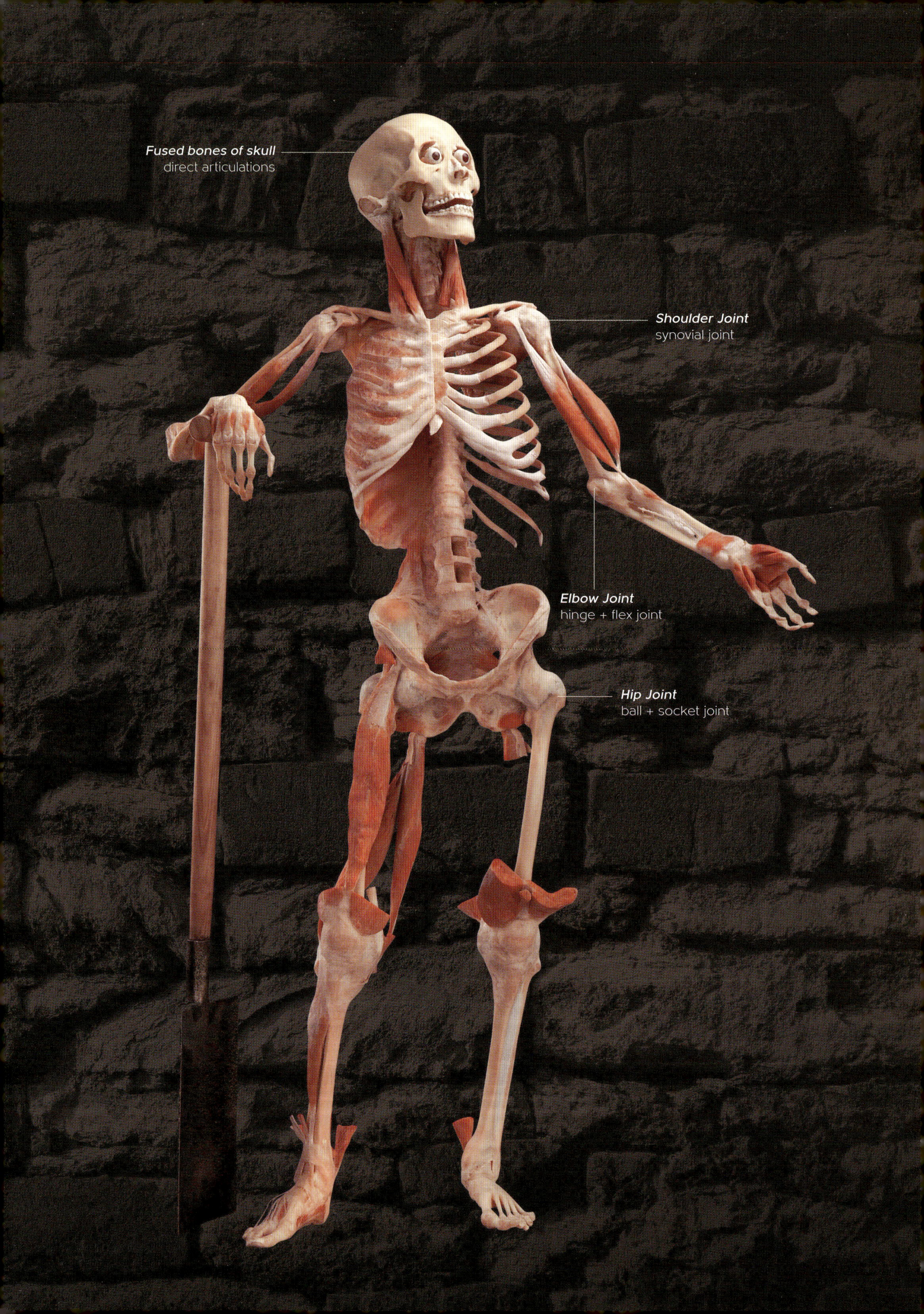
Fused bones of skull
direct articulations
Shoulder Joint
synovial joint
Elbow Joint
hinge + flex joint
Hip Joint
ball + socket joint

ARTICULAR SPECIMEN

THE BONES, JOINTS, AND MUSCLES IN THE BODY COMBINE TO FORM THE LOCOMOTIVE SYSTEM.

The bones act as levers that are moved (articulated) at the joints by the contraction and release of the skeletal muscles. These three groups coordinate to produce the complex movements the body undertakes with each moment. Even reading this booklet requires a tremendous amount of coordination between your bones, joints, and muscles.

There are two kinds of articulations (joints) in the body: direct articulations are firm and immovable joints such as the fusion of bones that creates the skull; synovial joints such as the wrist and knee joints allow for flexible movement. Most joints in the human body are synovial joints.

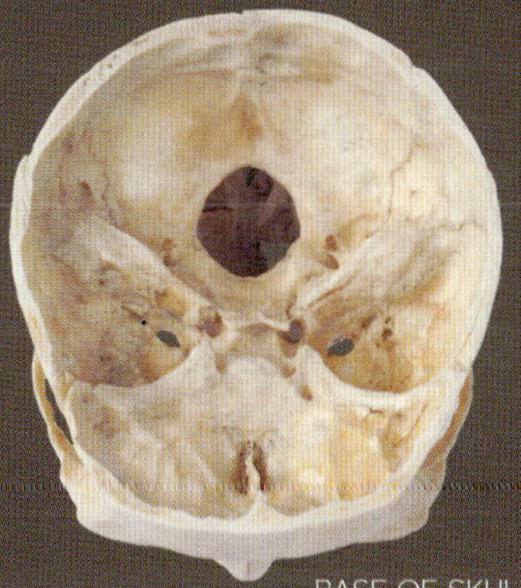

BASE OF SKULL (INTERIOR)

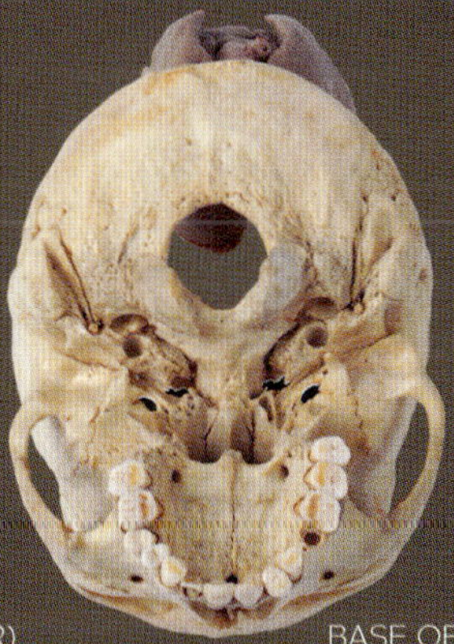

BASE OF SKULL (EXTERIOR)

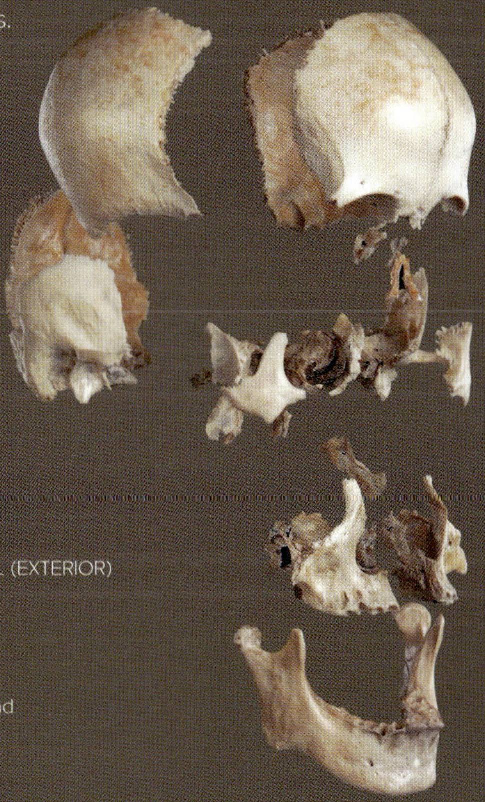

DISARTICULATED SKULL
The human skull is made of 23 bones. 8 cranial bones bones enclose the brain while 15 facial bones form the bony framework of the face. Responsible for the protection and support of the brain and the organs of sight, hearing and balance, these bones are closely fitted together by sutures or cartilage. The only movable bone of the skull is the lower jaw (mandible).

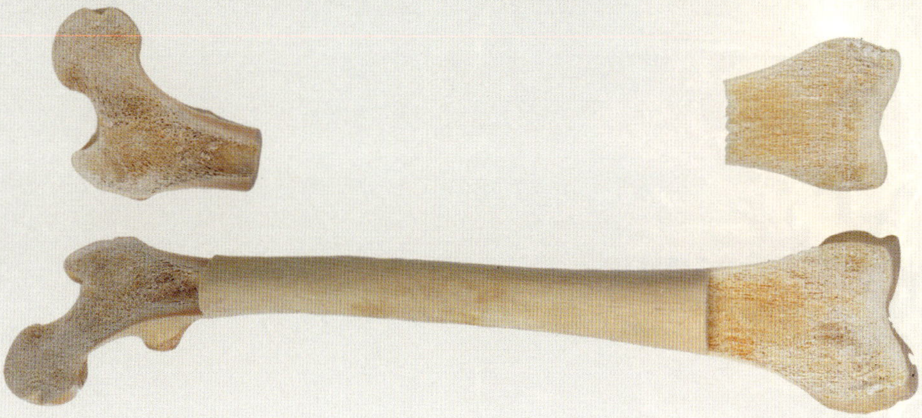

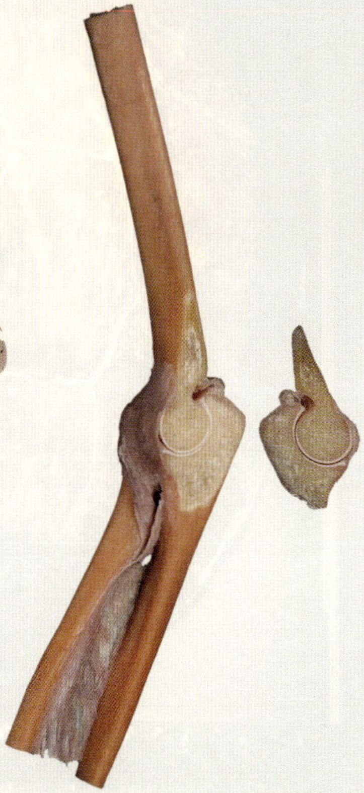

FEMUR SECTION

Bone contains both compact bone and spongy bone. Compact bone forms the shaft of the long bone and the external layer of other bones, allowing for resistance to stress. Spongy bone, found in the interior forms an expanded sponge-like mesh.

ELBOW JOINT

Made up of the humerus, radius and ulna, the elbow makes two simple movements: flexion and extension.

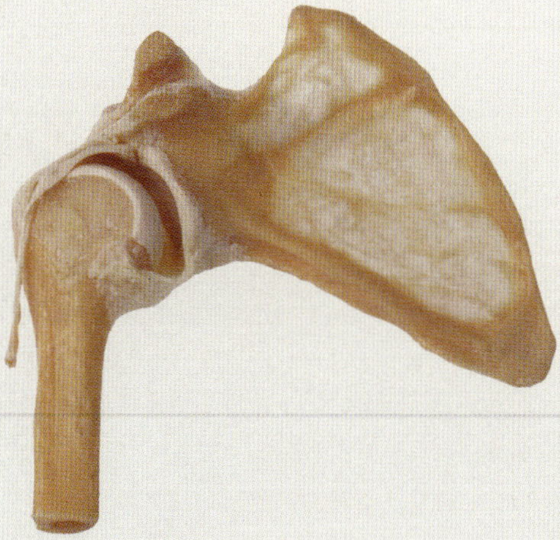

SHOULDER JOINT WITH OPEN SYNOVIAL CAPSULE

The shoulder joint is formed by the head of the humerus and the glenoid cavity of the scapula. Its structure enhances the extent of movements, but decreases the stability of the joint, thus making dislocation of the shoulder joint more likely.

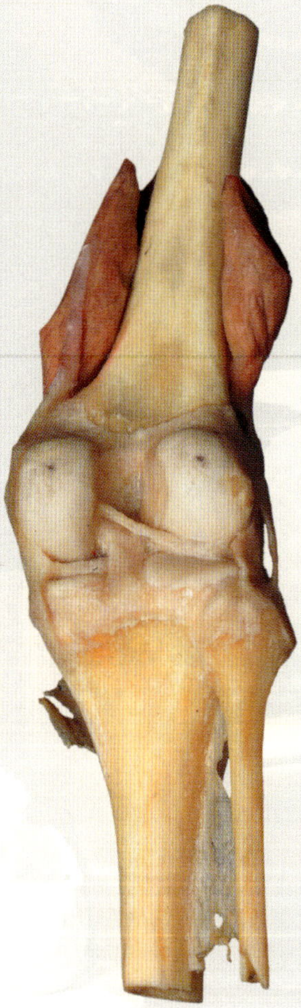

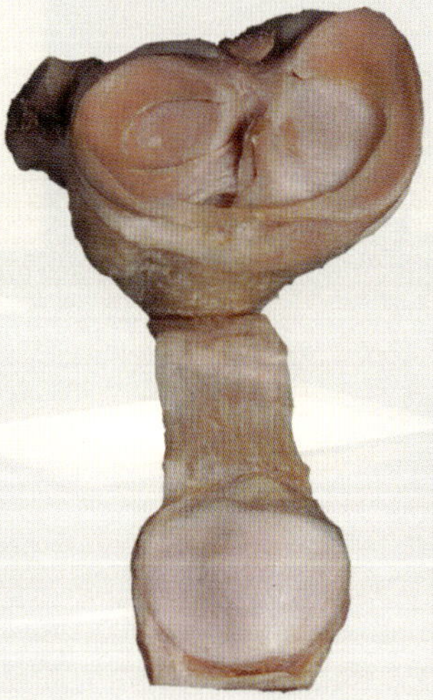

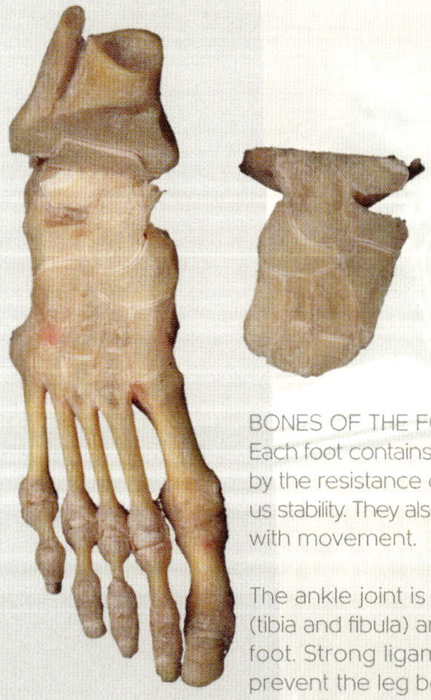

BONES OF THE FOOT AND ANKLE

Each foot contains 26 bones. The arches, formed by the resistance of muscles and ligaments, give us stability. They also protect us by reducing impact with movement.

The ankle joint is formed by the two leg bones (tibia and fibula) and the uppermost bone of the foot. Strong ligaments connect the bones to prevent the leg bones from slipping.

KNEE JOINT

The knee bears most of the body's weight. A flexible joint, it is reinforced by several ligaments such as the anterior and posterior cruciate ligaments (within the capsule) and the tibial and fibular collateral ligaments (along each side) to reinforce its stability.

THE MENISCI

Two crescent fibrocartilagenous plates in the cavity of the knee joint, the menisci, act as "resilient pads" to absorb shock and deepen the articular surfaces of the tibia.

VERTEBRAL COLUMN

The vertebral column consists of 24 individual vertebral sections, which include the cervical, thoracic, and lumbar vertebrae and the fused sacrum. The spinal cord, the spinal nerve roots and their coverings are located within the vertebral canal. The adult vertebral column has four physiological curves, which act as a "spring" making the column more elastic.

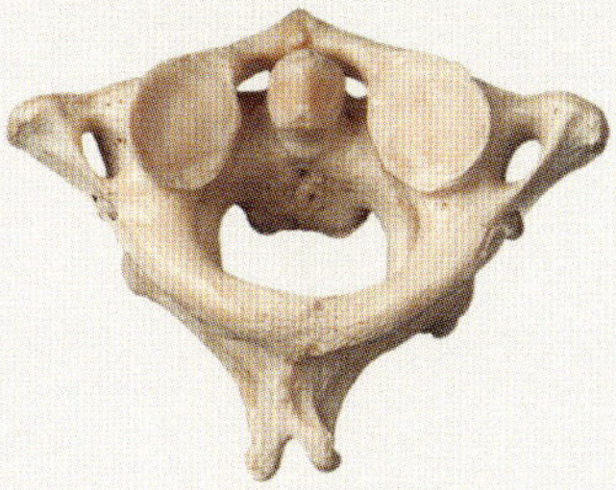

ATLAS + AXIS

The atlas (the first cervical vertebra) connects the skull and the axis (the second cervical vertebra), allowing for support and rotation of the skull.

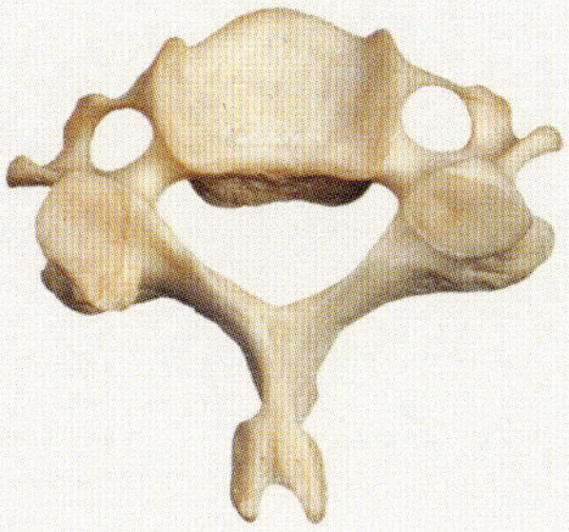

CERVICAL VERTEBRA

The seven cervical vertebrae provide support and stability to the neck. Among the smallest spinal vertebrate, they have the highest ability of movement.

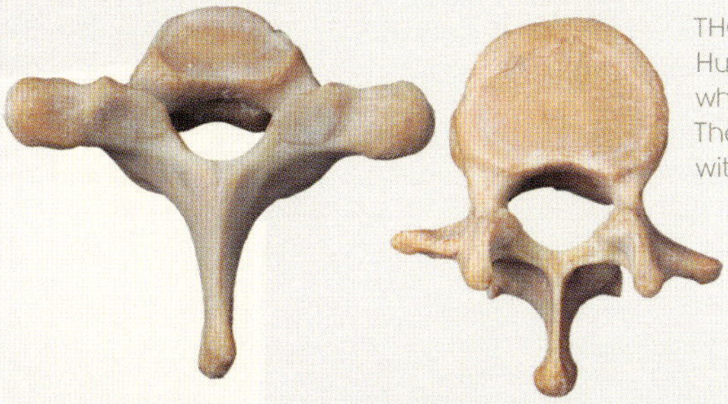

THORACIC VERTEBRAE

Humans have 12 thoracic vertebrae, which increase in size as they descend. Their shape allows them to connect with the ribs.

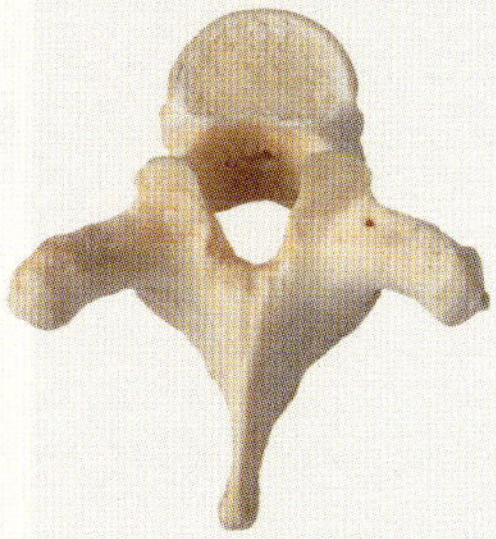

LUMBAR VERTEBRA

Humans have five lumbar vertebrae. They are the thickest and strongest vertebrae as they provide significant support to the upper body.

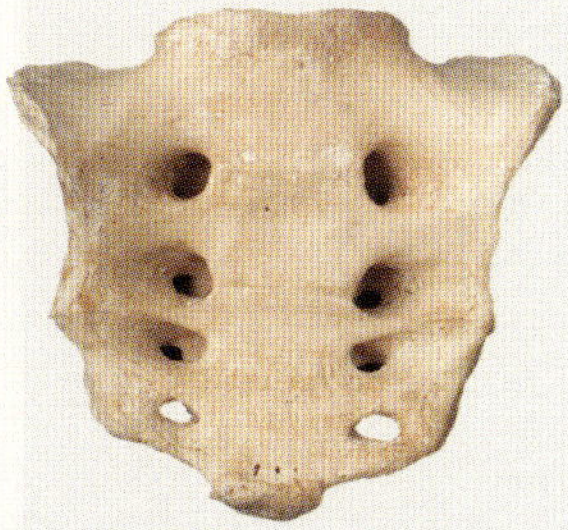

SACRUM

The sacrum consists of five fused vertebrae at the end of the vertebral column. Its name comes from the original Greek, which meant sacred bone because of the belief the soul resided within it.

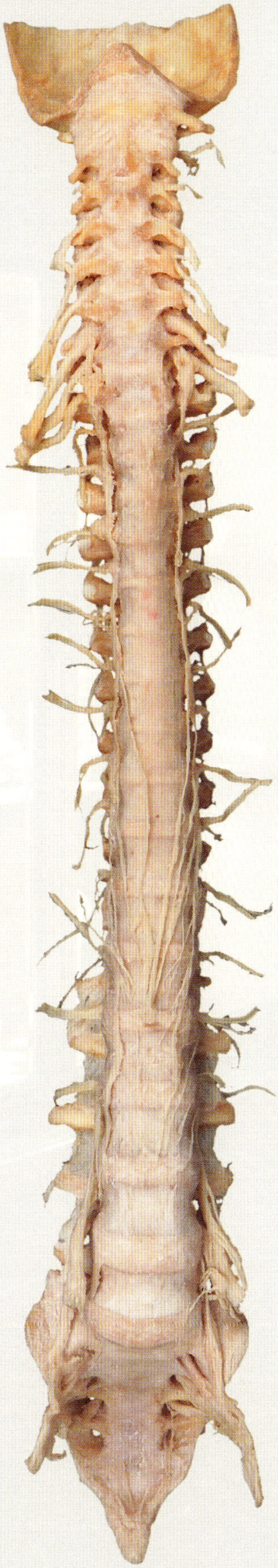

BREATHE

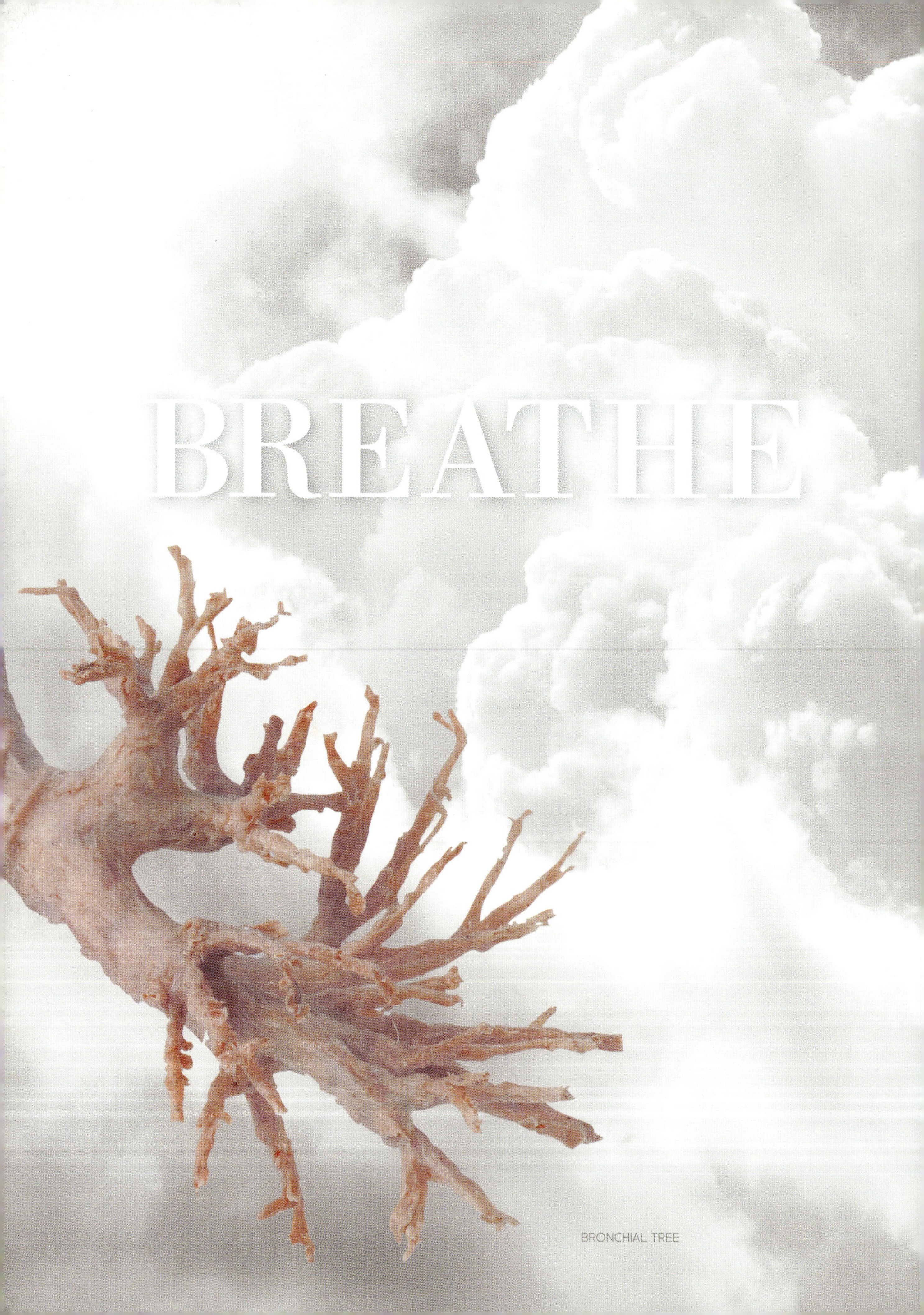

BRONCHIAL TREE

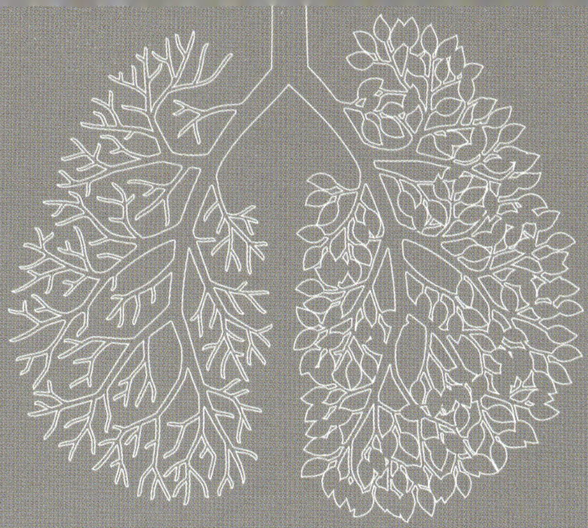

THE RESPIRATORY SYSTEM

24 hours a day, 7 days a week, the respiratory system delivers life-giving oxygen to awaiting blood cells and removes waste products from the body. Also a main component of speech and the human voice, this system ensures that we have the oxygen we need and enables our desire to express ourselves.

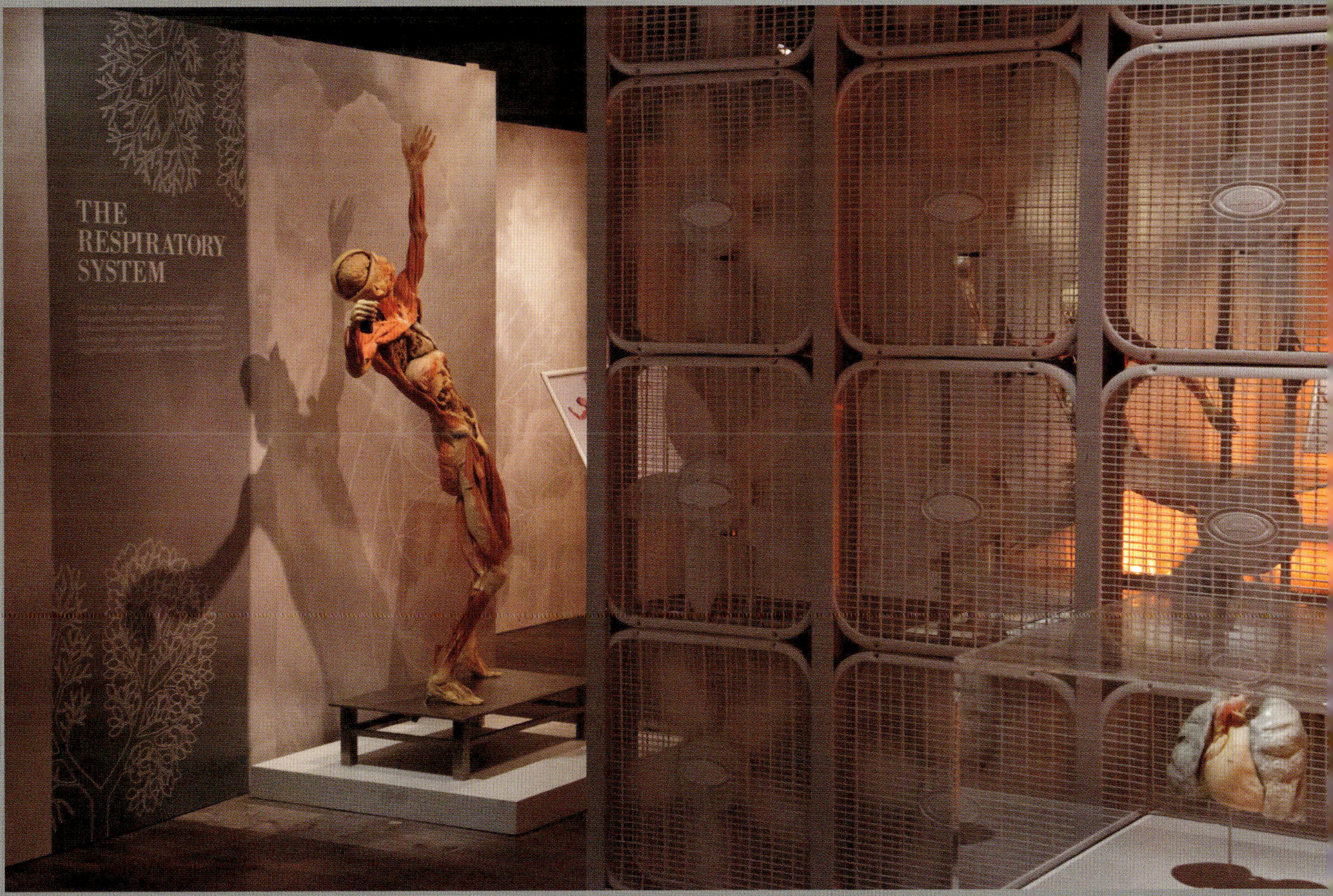

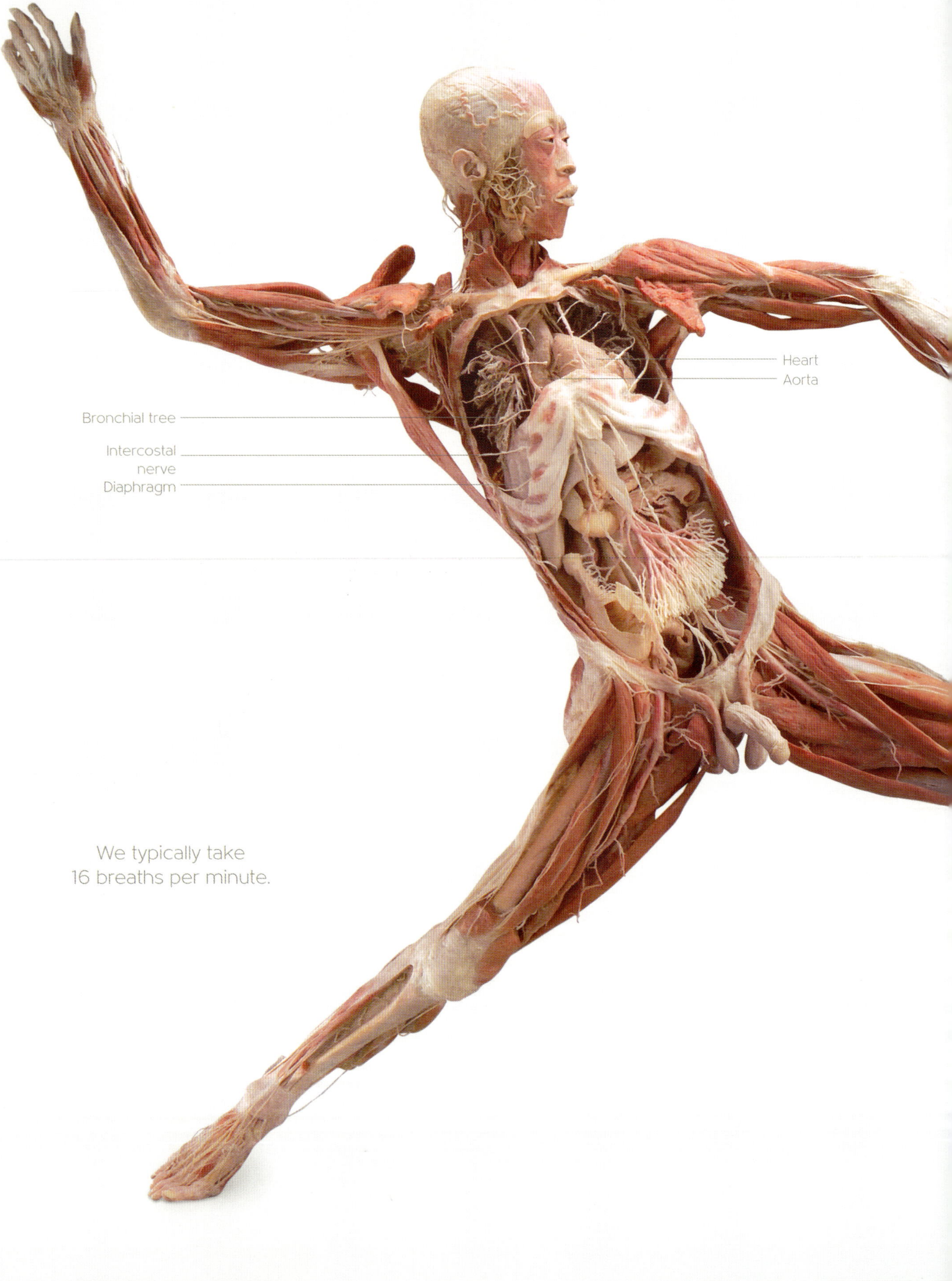

We typically take
16 breaths per minute.

RESPIRATION + CIRCULATION

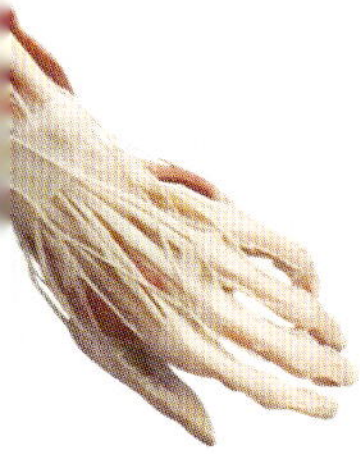

THIS DISSECTION REVEALS THE INNER WORKINGS OF THE CORE OF OUR RESPIRATORY SYSTEM AS WELL AS ITS IMPORTANCE TO CIRCULATION.

Here, the blood vessels that surround the bronchial tree are exposed, demonstrating the significant amount of blood vessels that operate in this region. They both absorb oxygen from the lungs and dispel cellular waste such as carbon dioxide into the lungs for exhalation from the body. This gas exchange occurs in the deepest branches of the bronchial tree (the alveoli) where a thin, semi-porous membrane divides the lung wall from the bloodstream. Here oxygen transforms from a gas into a solution and enters the bloodstream. Red blood cells contain the protein hemoglobin, which bonds to oxygen and pulls it from the solution. As hemoglobin forms a stronger bond with oxygen than carbon dioxide, any carbon dioxide carried by the hemoglobin is released and passes in the opposite direction through the membrane.

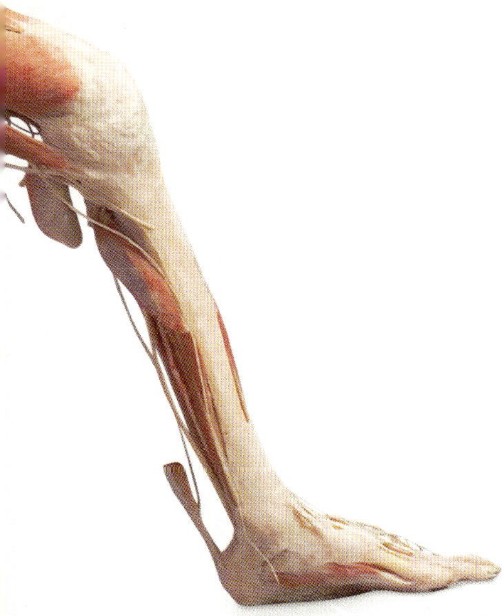

When hemoglobin is fully saturated with oxygen, it becomes bright red in color, which gives blood its unmistakable appearance. Once in the blood stream, red blood cells deliver oxygen to every cell in the body. Oxygen is essential for cells to function, providing the fuel that allows our cells to perform their life-sustaining tasks.

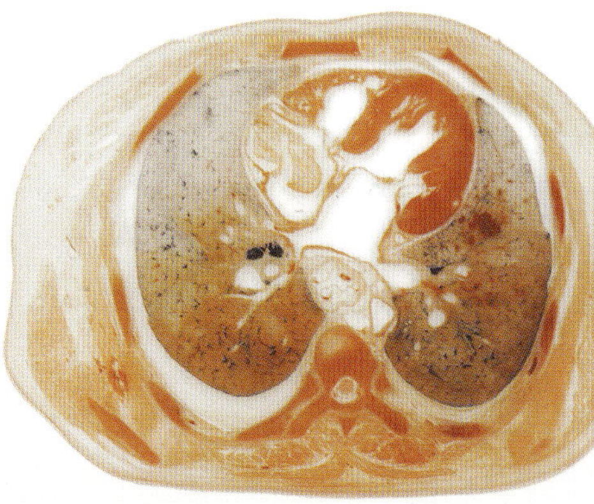

Transverse section of the chest reveals the close relationship between the heart and lungs.

> When people ask me what the most important thing is in life, I answer: 'Just **breathe**'.
>
> Yoko Ono

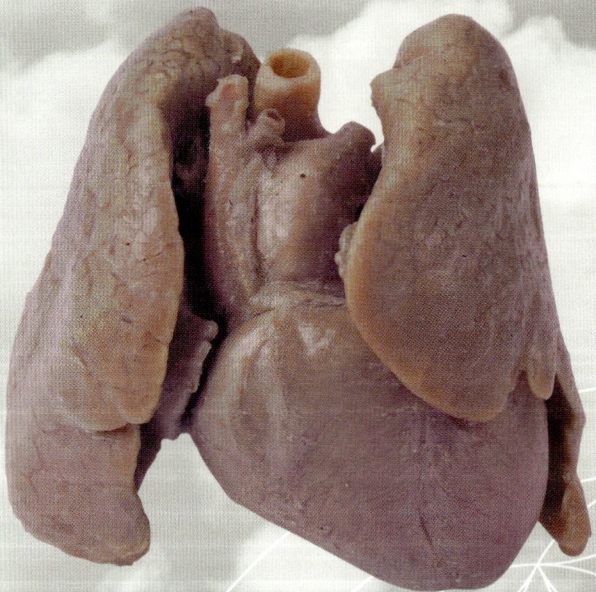

FETAL LUNGS
The fetus does not use its lungs while in the womb, however, closer to birth, it does "practice" breathe by circulating amniotic fluid in and out of the mouth.

TO BREATHE TO LIVE

Life begins with a breath - a sudden gulp of air into the lungs followed by the plaintive cry of the newborn, announcing "Here I am. I am ready." And so it begins, a lifetime of breaths—more than half a billion.

Inhale. Exhale. Repeat.

The gentle breathing of the sleeping child. The rapid puffs of the teenager prankster who scampers from the scene. The hard draws of the victor and the vanquished. The purposeful focus of the Yogi. The inaudible whistle of your love late at night. The deliberate pull through the nostrils to take it all in.

Inhale. Exhale. Repeat.

Beyond culture, beyond race, beyond religion, breath is undeniable. It is inspiration, a call to the spirit of life: "Come, fill these empty sponges." Shared by us all, breathing is willful and defiant, an act of the unconscious, that declares "I am!" through the cycle of years and minutes, renewing every four seconds, restoring us, remaking us, bringing new beginnings.

Inhale. Exhale. Repeat.

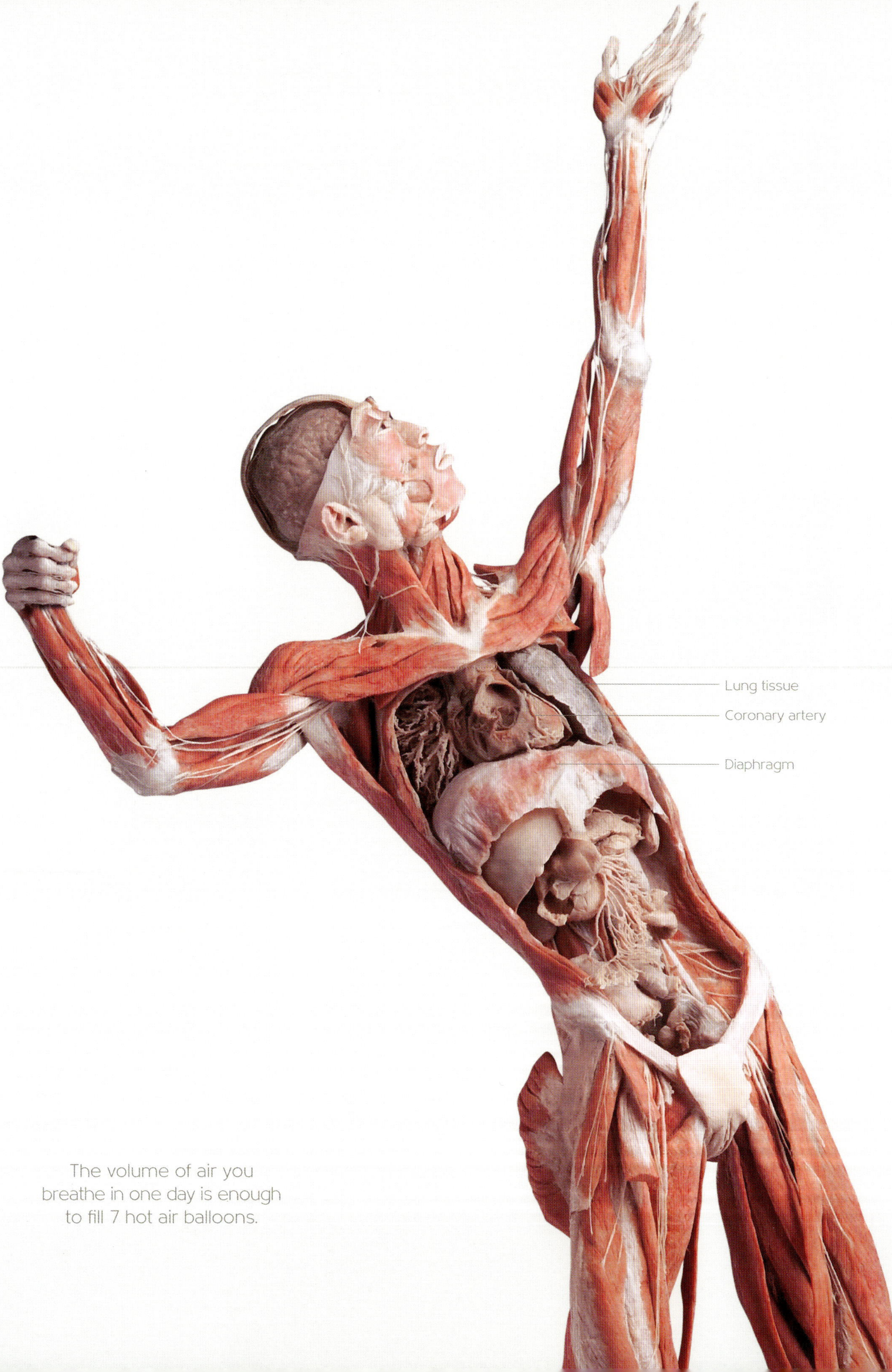

The volume of air you breathe in one day is enough to fill 7 hot air balloons.

AIR SUPPORT

While the process of blood absorption occurs deep within the lungs, the respiratory system includes a series of supporting passageways and tissues that allow the air we breathe to reach the areas of gas exchange. Tiny hairs called cilia protect the nasal passageways and other parts of the respiratory tract, filtering out dust and other particles that enter the nose as we breathe. These particles are either exhaled or absorbed into the digestive tract.

The airways of the nose and mouth meet at the pharynx where air is directed to the trachea, which is held open by cartilage rings to ensure air passes into the lungs. The lungs also contain elastic tissues that allow them to inflate and deflate without losing shape and are encased by a thin lining called the pleura.

The lungs are put into action by the diaphragm. When we inhale, the diaphragm moves down expanding the size of the lungs, thus reducing their air pressure and allowing outside air to flow in. When we exhale, the diaphragm moves upward, forcing the chest cavity to contract and pushing air out of the lungs.

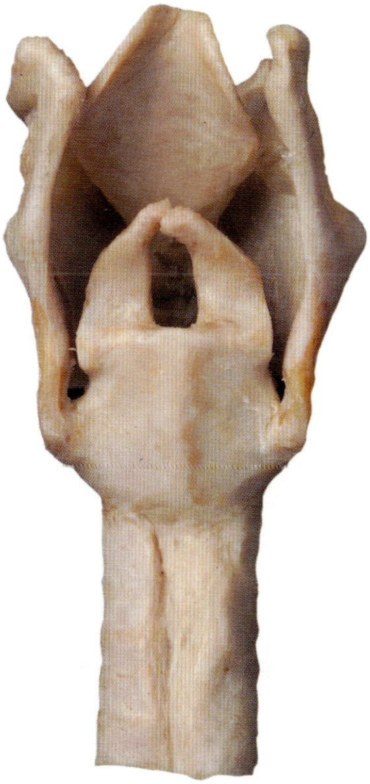

This view of the trachea displays the rigid nature of the airway, whose cartilage allows it to stay open for breathing. The flap at the top, the epiglottis, closes while swallowing food or water to prevent the movement of food or beverage into the airway.

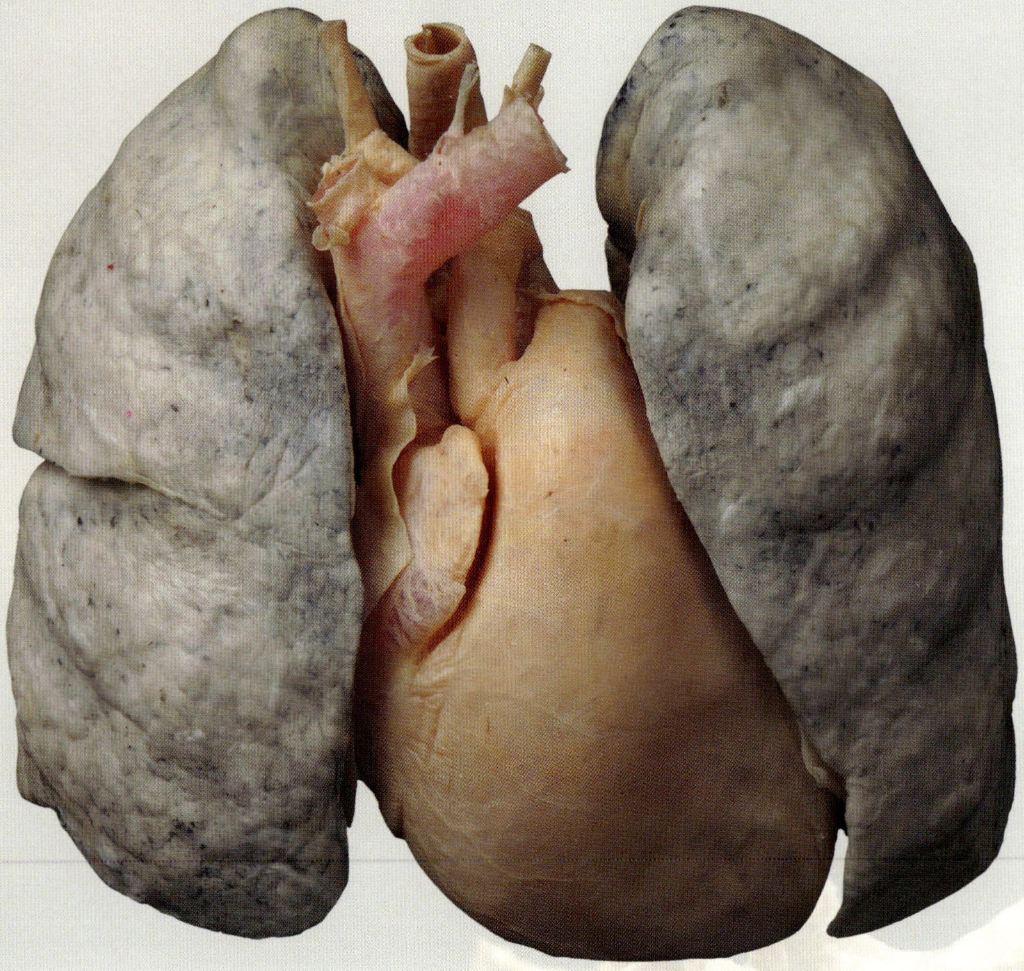

NORMAL LUNGS
The lungs are the essential respiratory organs, which are situated in the thoracic cavity and on both sides of the heart. Children's lungs are pink, but as we age dust such as carbon and other particles accumulate in the lungs. As a result, they appear grey in color.

The left and right lungs differ slightly in shape and size: the right lung has three lobes; the left lung has only two lobes to make room for the heart.

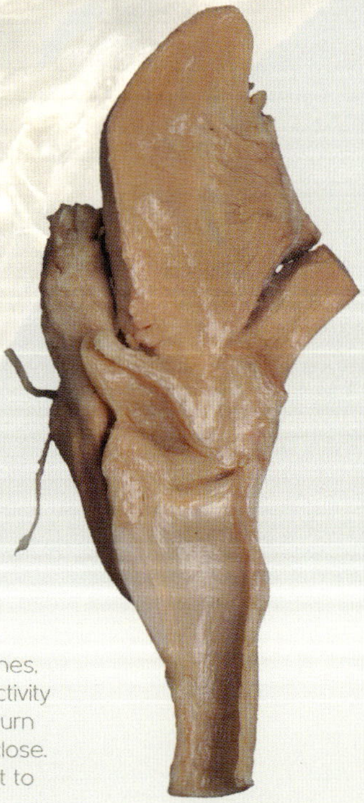

INSIDE THE LARYNX
The vocal cords, two folds of mucous membranes, are found within the laryngeal cavity. Muscular activity constricts or relaxes the vocal cords, which in turn cause the gap between the cords to widen or close. This gives us our voices, an important element to self-expression.

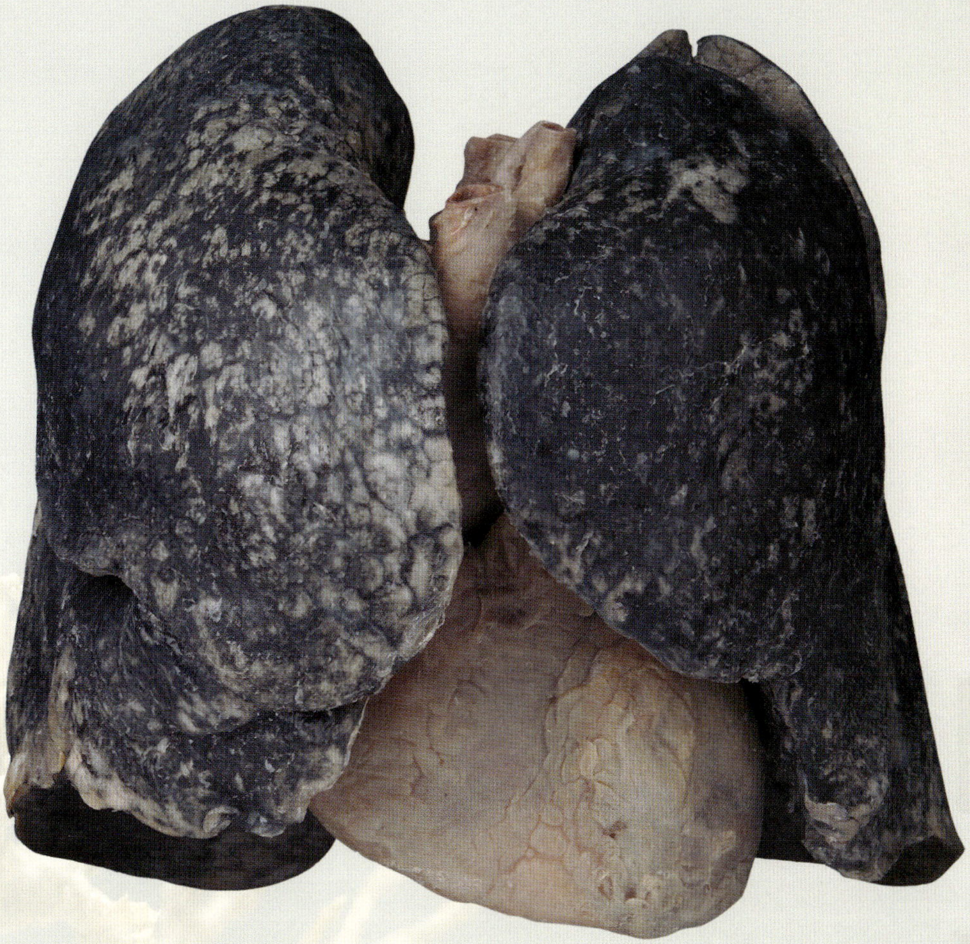

SMOKER'S LUNGS
These lungs appear black and have great discrepancy from the non-smoker's lungs. This blackening is most likely the result of tar from cigarette smoke and carbon particles from polluted air.

Cigarettes damage the tissues of the lungs and the airways. They increase the risk of lung cancer and heart disease.

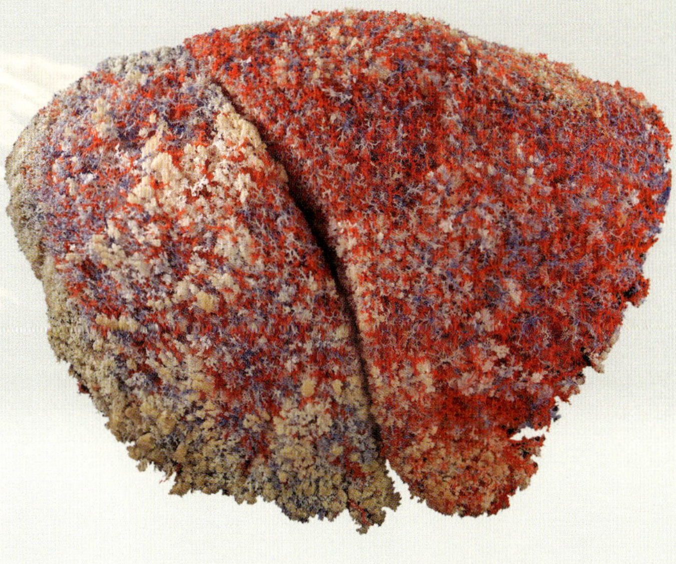

BLOOD VESSELS OF THE LUNG
When infused with a liquid polymer, which then hardens, the lung reveals its vast network of blood vessels that support lung tissue and lung function. Most importantly, these blood vessels serve as the site of gas exchange–in which oxygen in the air we breathe in is delivered to awaiting red blood cells and carbon dioxide is released.

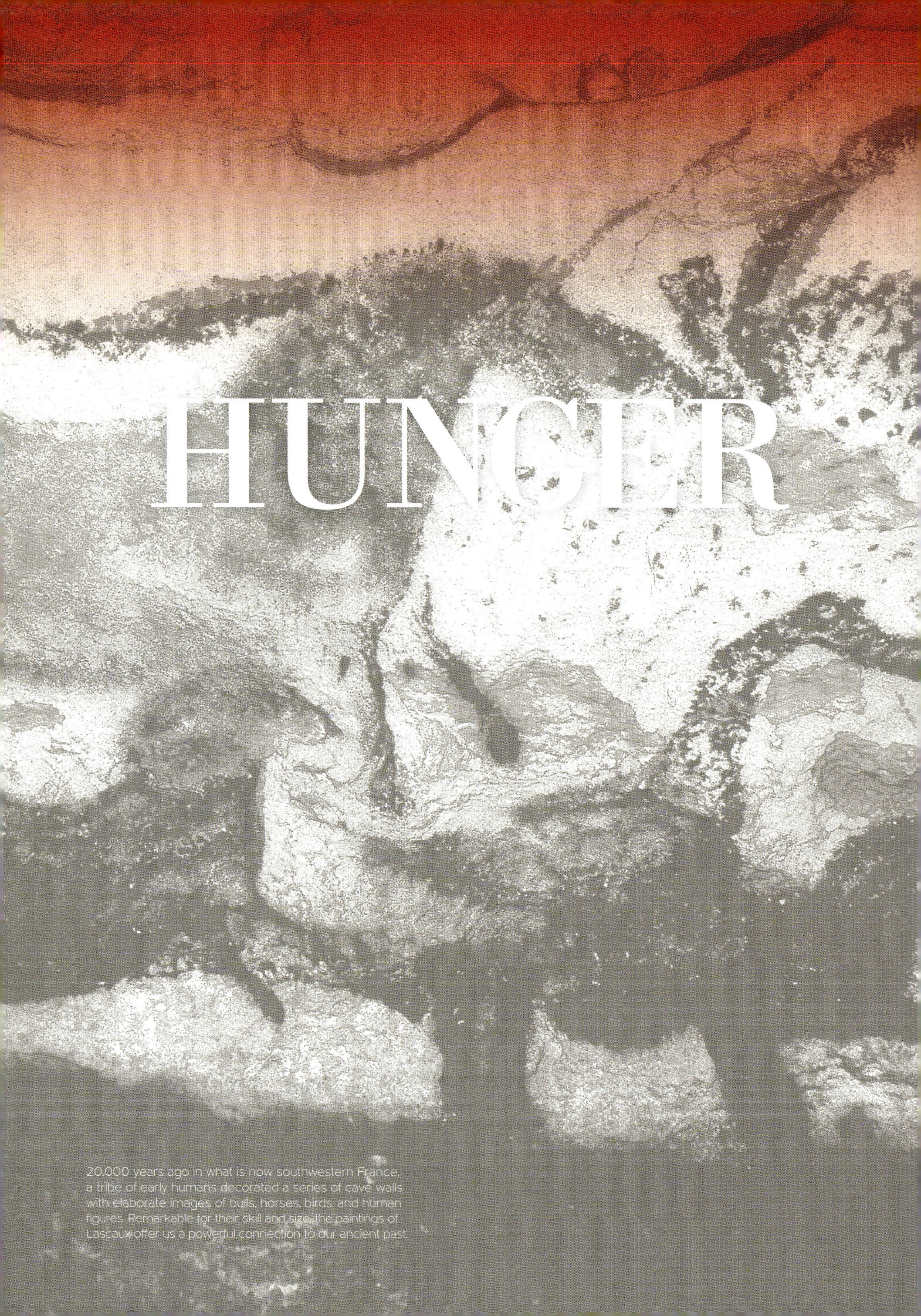

HUNGER

20,000 years ago in what is now southwestern France, a tribe of early humans decorated a series of cave walls with elaborate images of bulls, horses, birds, and human figures. Remarkable for their skill and size, the paintings of Lascaux offer us a powerful connection to our ancient past.

THE DIGESTIVE SYSTEM

Essential to life, the digestive system converts the foods we eat into the fuel we need to survive and thrive. More than thirty feet (9.1 m) in length, the digestive system breaks down foods and liquids into digestible components, absorbs nutrients that the body converts to energy for the cells, and compacts and eliminates wastes. Packed with nerves, the digestive system is also often thought responsible for those "Gut Feelings."

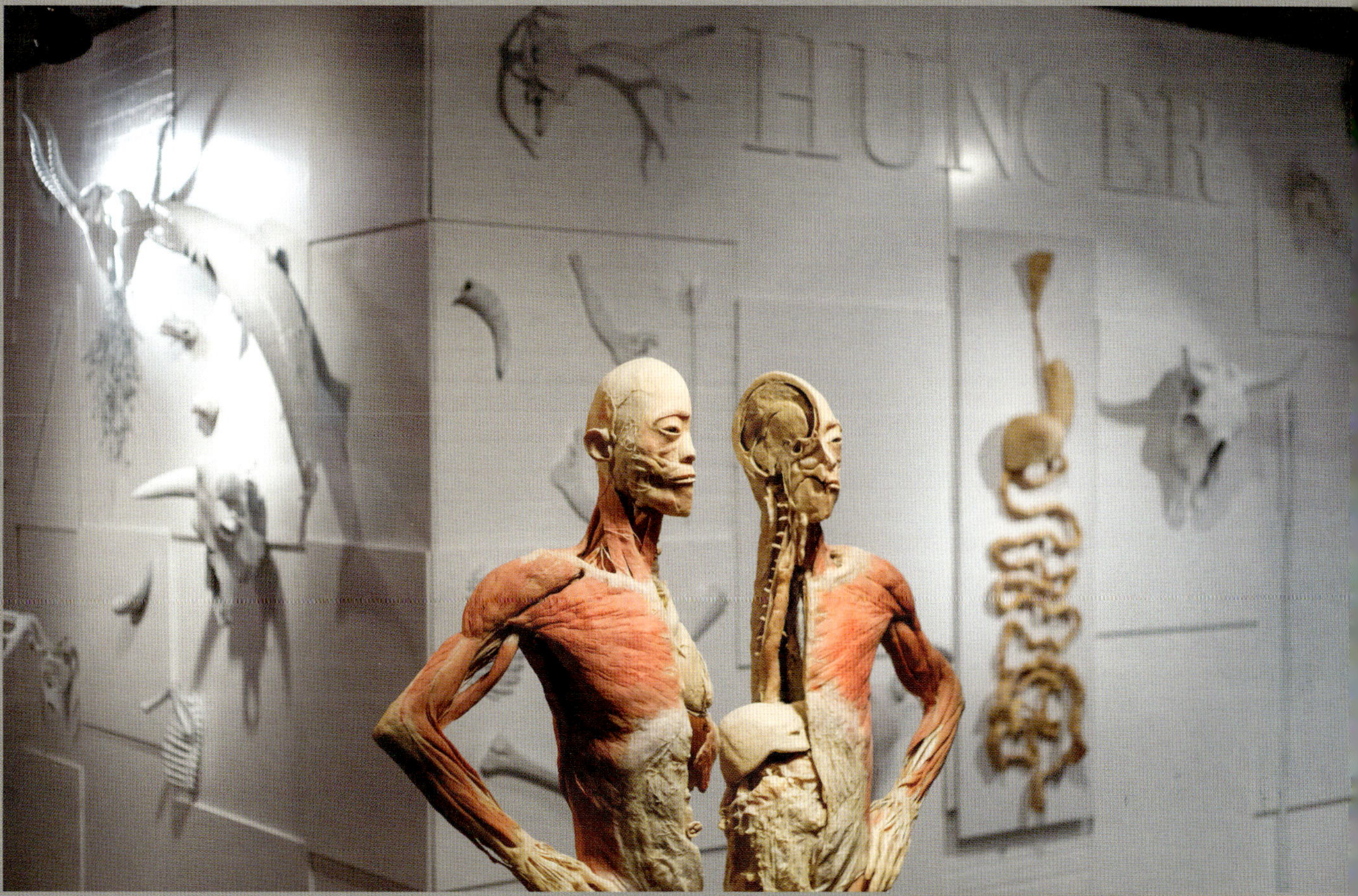

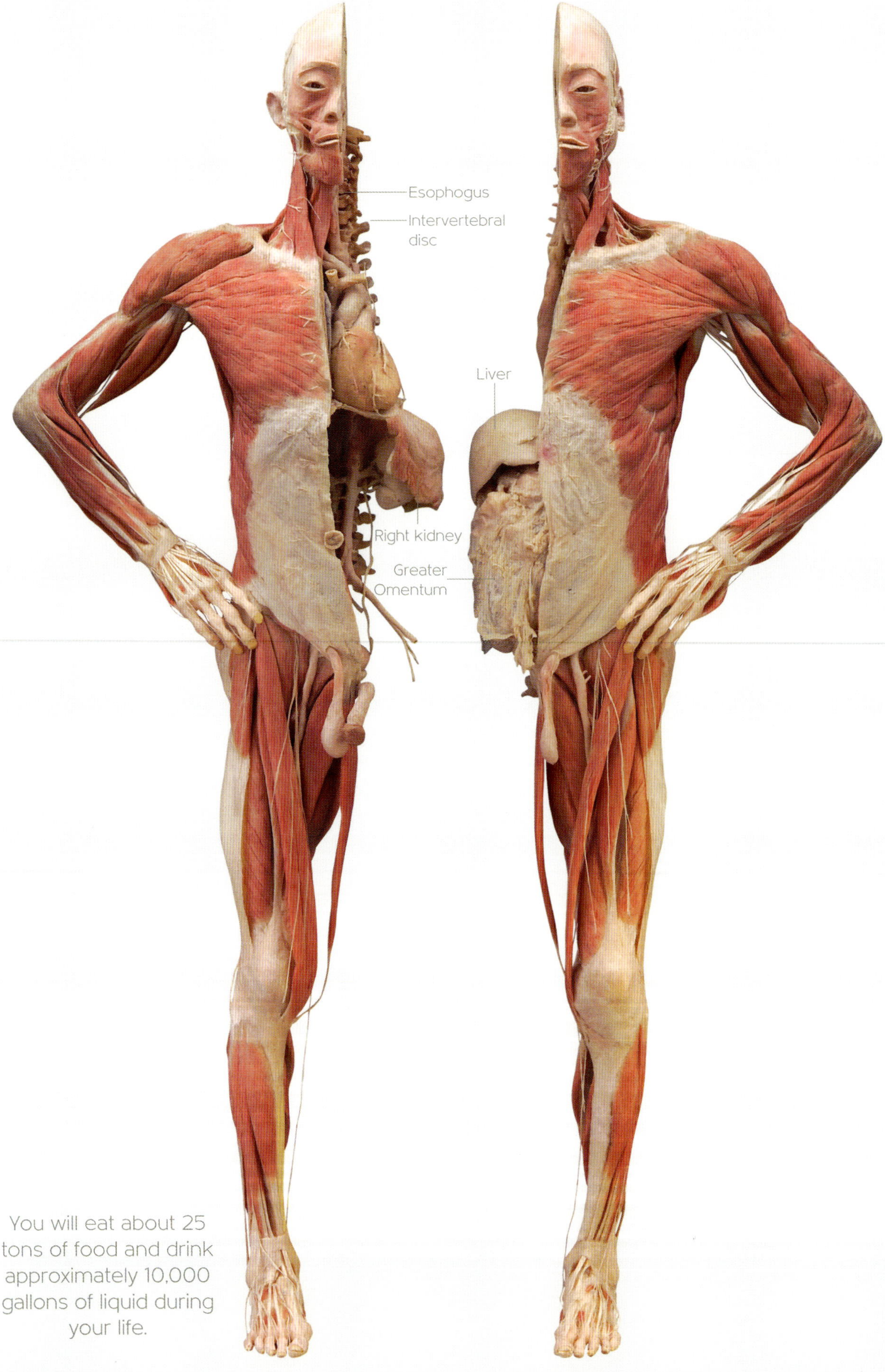

You will eat about 25 tons of food and drink approximately 10,000 gallons of liquid during your life.

MEDIAN SAGITTAL LINE

THIS SPECIMEN WAS DISSECTED ALONG THE MEDIAN-SAGITTAL LINE, CREATING A REVEALING VIEW INTO THE VISCERA, AND HEAD AND NECK REGIONS.

The relationship of the organs within the viscera displays a compact and efficient arrangement. The diaphragm in the middle divides the viscera into two cavities–the thoracic cavity and abdominal cavity. The heart, located in the thoracic cavity, lies in close proximity to the lungs and the brain. In this way it can quickly receive oxygenated blood from the lungs and send deoxygenated blood to them to expel waste products; and it can keep a steady flow of blood towards the hungry brain.

The organs of digestion, reproduction, and waste processing are located in the abdominal cavity. The liver is very prominent in this dissection and is one of the largest organs in the body due to its role in cleaning the blood, creating bile for digestion and creating vitamins. The liver sits atop the stomach and digestive organs, and is in close proximity to the kidneys, which also share a responsibility in filtering the blood.

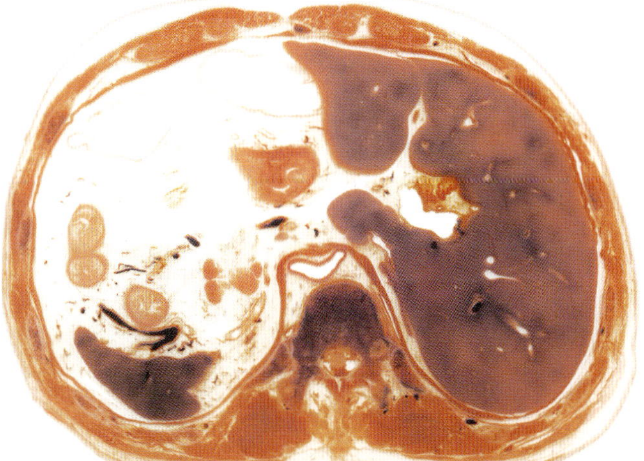

Transverse section of the abdomen reveals the relatively massive size of the liver (right), as well as the significant amount of open space dedicated to the intestinal tract.

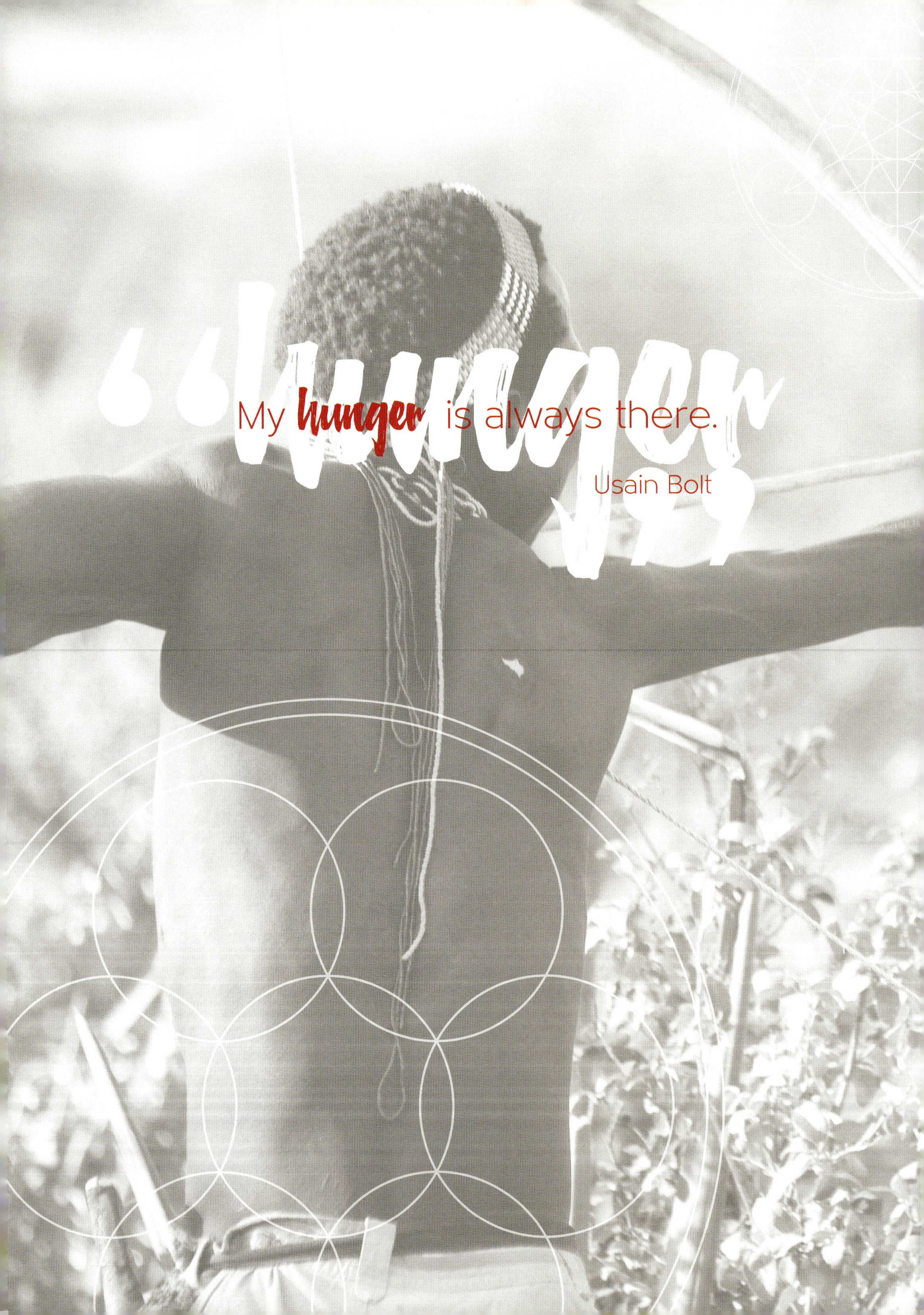
"hunger"
My hunger is always there.
Usain Bolt

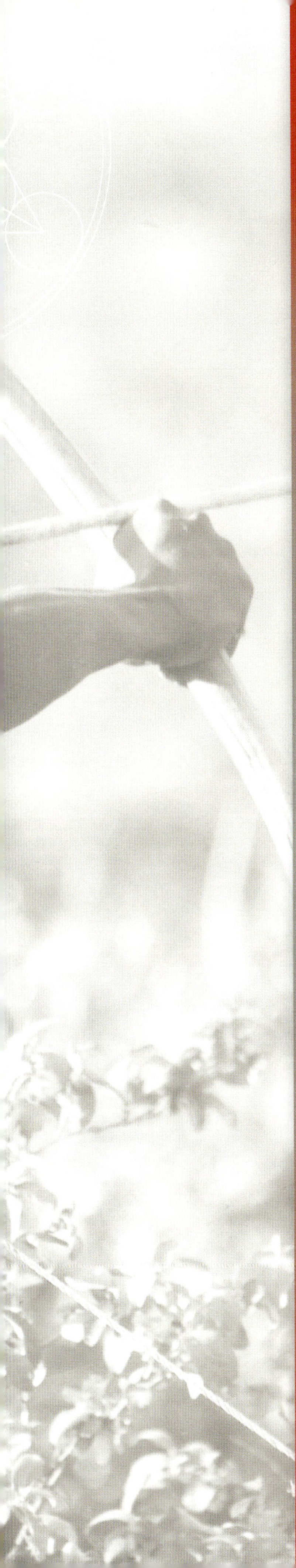

HUNT AND GATHER

For most of us, access to food is a certainty. We are lucky. We go to the shops or the restaurant.

For our first human ancestors, who appeared approximately 1.8 million years ago, finding food was the top daily priority. They ate what they found: plants, nuts, berries, shrubs. The quest for food was a way of life. Modern humans, who appeared about 200,000 years ago, followed in that same tradition. Eventually, with the development of nets and sharp implements, hunting and fishing became another way to find sustenance.

As the glaciers of the last Ice Age receded, tribes in temperate regions of the world were more inclined to remain in one place and started cultivating crops (first in Jericho about 8,000 years ago) and raising livestock (sheep and goats were first raised in the Middle East in what is now Iraq; pigs were most likely first raised in China). The predictable and dependable access to food changed our destiny–we stopped moving and settled. We put down roots and built up cities.

Our way of life today has come a long way from that of our ancestors who hunted and gathered, but their opportunistic imprint still lies deep within us. We feel it in the thrill of the chase, the pride in home grown vegetables, the visceral attraction and repulsion at the smell of blood in the butcher's shop.

Easy access to food, however, is still not everyone's reality in the 21st century. Poverty and hunger still blights too much of the world. Perhaps our innate instinct to generate "plenty" can be harnessed to correct this.

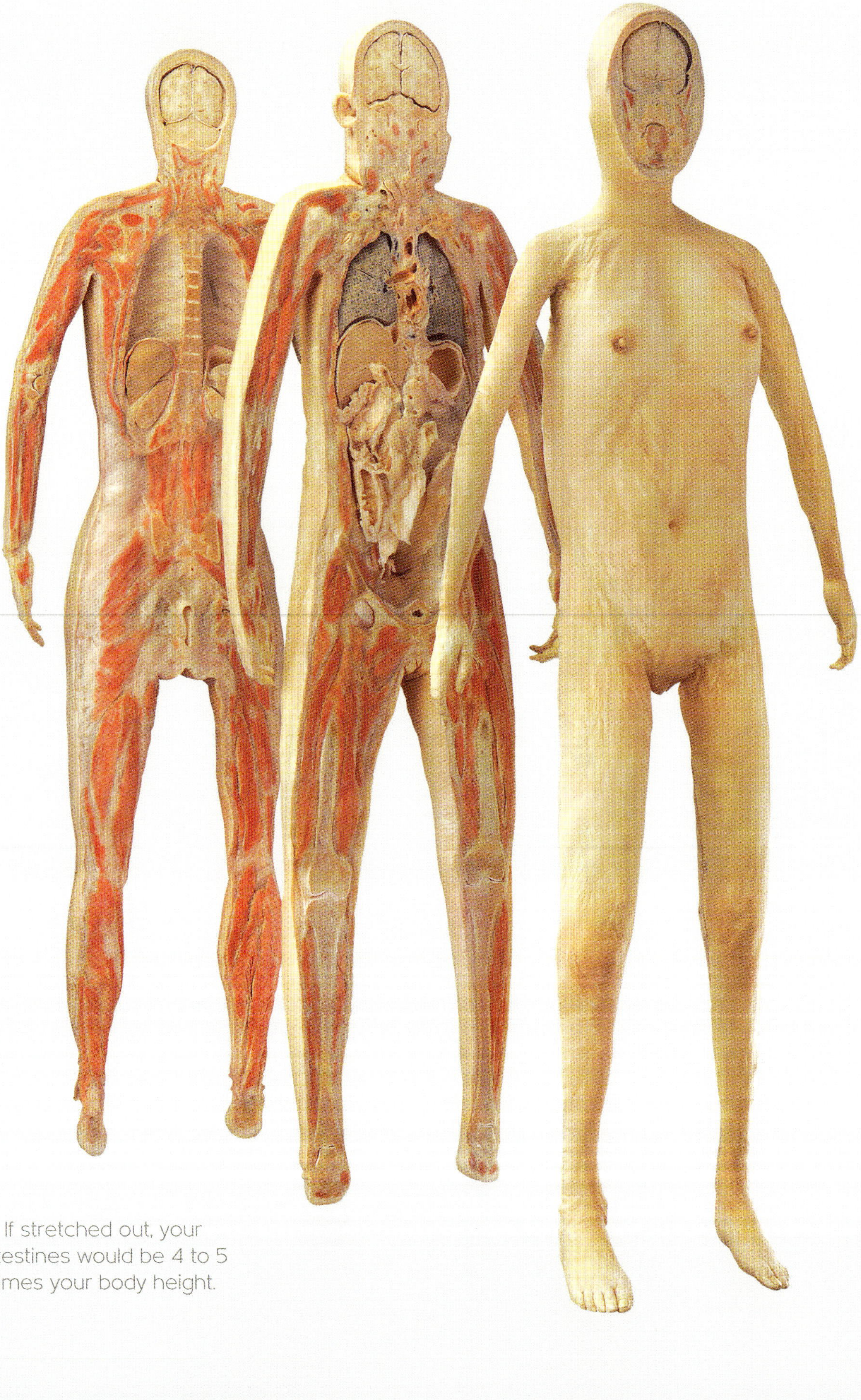

If stretched out, your intestines would be 4 to 5 times your body height.

BODY ARMOR

This specimen shows a female body in a series of coronal sections. Understanding how the structures of the body relate to one another in a sectional and regional manner is an important element in the training of physicians, especially those who study radiology and ultrasonography.

Here, we also have an opportunity to understand how the musculature and skeleton acts as a type of "armor" for the body–protecting its internal organs that are vital to fuel the muscles and skeleton and carry on the life of our species. The foremost section of this specimen mainly displays skeletal muscle, ligaments and bones. The second section displays the internal organs including a fascinating view of the brain. The third section of this specimen displays how the skeleton and musculature also provide protection from behind; this section also features an exciting dissection of the spinal column revealing the intervertebral discs.

In addition, this specimen helps us understand how adipose tissue is distributed throughout the body. For women, adipose tissues typically gather around the mammary glands, lower abdomen, the back of the legs and the gluteal and hip regions. While much is made today about fat, it is important for us to remember that we cannot live without fat. It absorbs nutrients, helps maintain body temperature and is an important energy reserve.

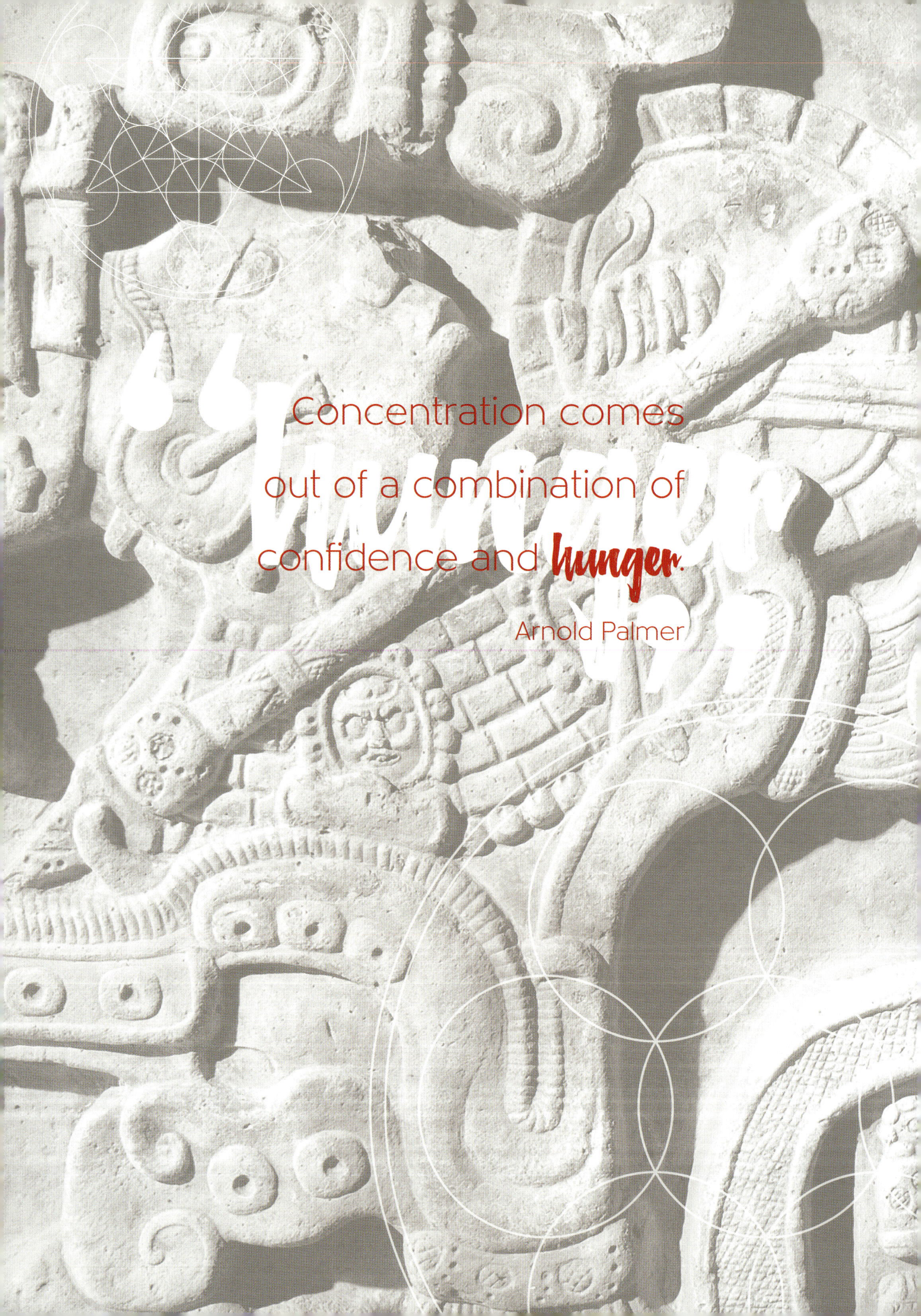

> Concentration comes out of a combination of confidence and *hunger*.
>
> Arnold Palmer

FAMILY MEAL

For what we are about to receive...

It may be a hut, home, or banquet hall. The table is smooth dirt, rough wood, or fine linen. Knives, forks, spoons, chopsticks, skewers, tongs, fingers. Evening meal. Sunday lunch. Christmas dinner, birthday tea, Easter lamb, Passover feast.

Together we sit and eat. We break bread together to celebrate. We raise our glasses to the moment at hand, to our host, our honored guests, to each other.

Yet we are toasting others–our ancestors who brought us to this moment, who tamed the land, who gathered the herds, who harnessed fire from the sky. Their instincts to harvest and hunt in the times before history make our lives possible.

Each meal stirs in us a shadowy memory of an ancient victory, a brief triumph over hunger–as essential and primordial as the world itself.

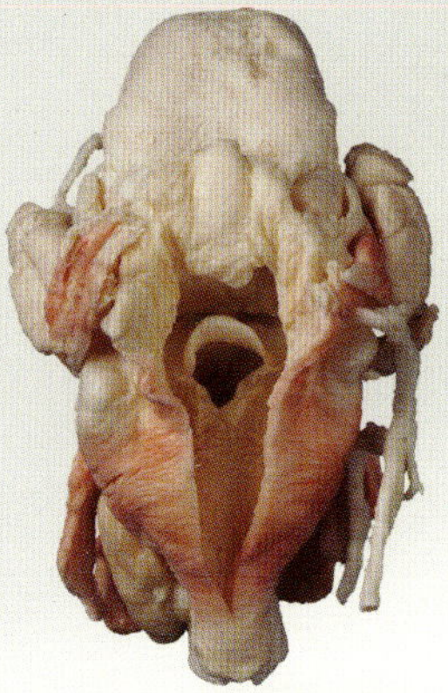

TONGUE

The tongue is a muscular organ in the mouth that assists with chewing, swallowing, and taste. The tongue's texture is covered with tiny bumps called papillae, which are the locations of thousands of taste buds. The taste buds send information to the brain of the four primary sensations: sweet, sour, bitter, and salty. Recently a fifth taste has been identified called umami. Borrowed from a Japanese word, umami is best translated as "satisfying savory taste" such as the ones delivered by meat broths, mushrooms, and earthy greens.

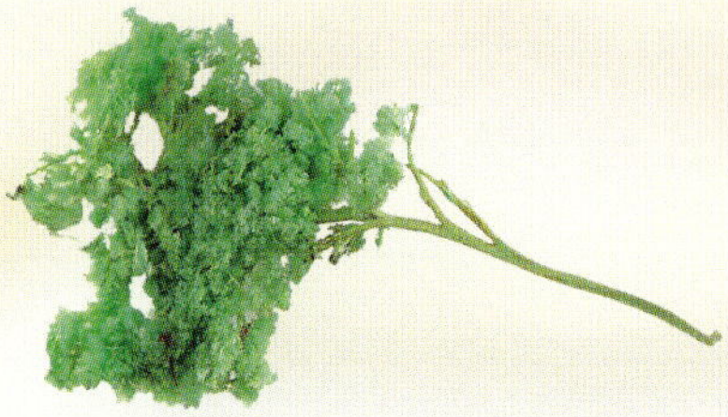

DUCTS OF PAROTID GLAND

The major salivary glands in humans, the parotid glands are located on either side of the mouth and in front of both ears. The largest of the salivary glands, the parotid secretes saliva through the parotid duct into the mouth to assist with chewing and swallowing while also beginning the breakdown of starch.

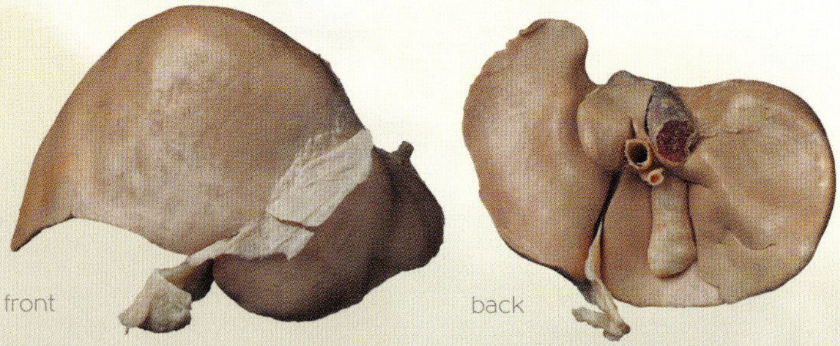

LIVER

The largest gland in the body, the liver is the most active organ in metabolism. It participates in the synthesis, conversion and decomposition of proteins, fats, and carbohydrates while also producing vitamins. Here toxins are also broken down and digestive bile is produced.

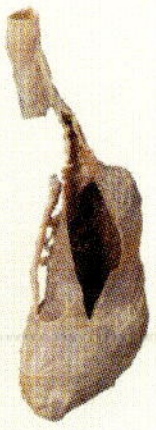

GALL BLADDER

The gall bladder stores bile created by the liver until it is required for digestion.

SMALL INTESTINE

On average 10 feet in length, the small intestine digests the food we eat into nutrients and minerals that the body can use for fuel. Here proteins, lipids, and carbohydrates are converted into molecules that can be assimilated into the bloodstream and put to use. Its many interior folds greatly increase its surface area allowing for hundreds of millions of tiny, finger-like projections called villi to grab onto the digested molecules and draw them into the bloodstream.

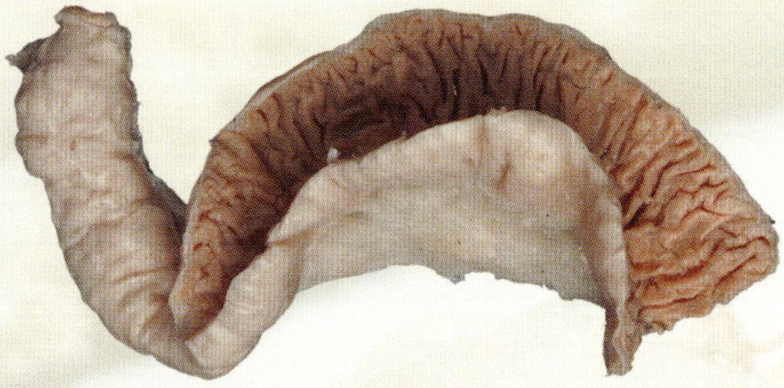

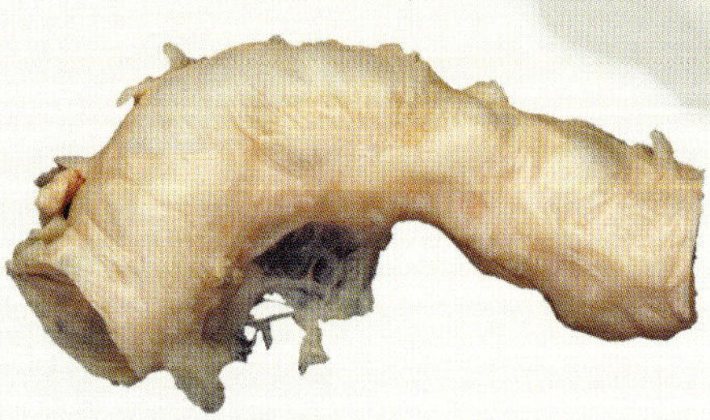

LARGE INTESTINE

The final section of the gastrointestinal tract, the large intestine absorbs water and vitamins while converting digested food into feces. About five feet in length, its walls are much thicker than those of the small intestine. The six sections of the large intestine are: the cecum; ascending colon; transverse colon; descending colon; sigmoid colon; and rectum.

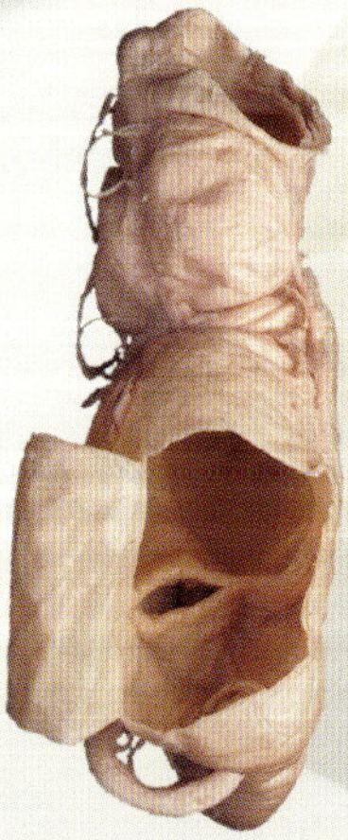

CECUM AND VERMIFORM APPENDIX

The cecum is the beginning portion of the large intestine. The vermiform appendix is a narrow, worm-shaped tube, which is about 9 cm long. Considered an evolutionary relic, the appendix is believed to have been used when our ancestors ate difficult to digest foods.

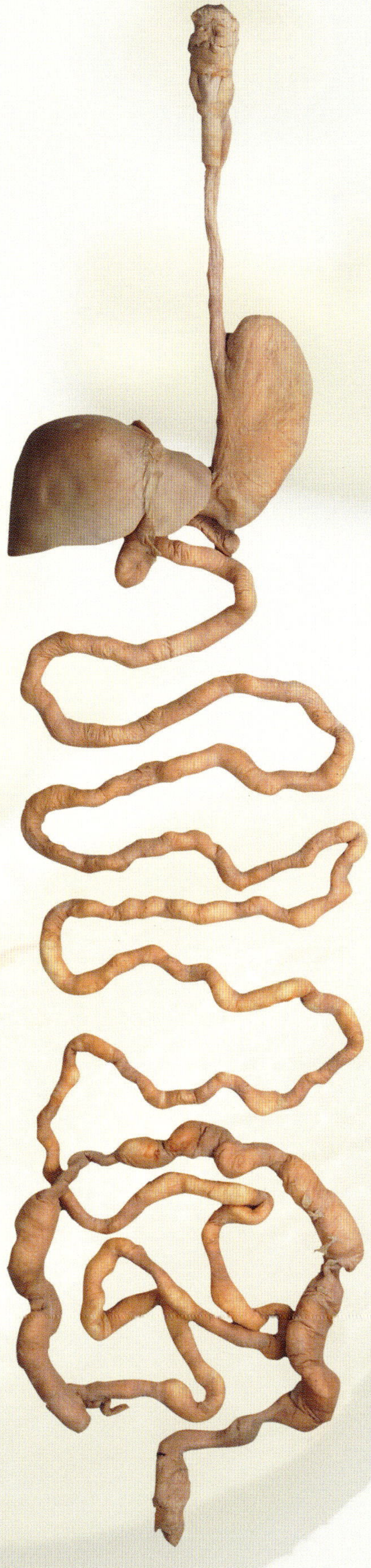

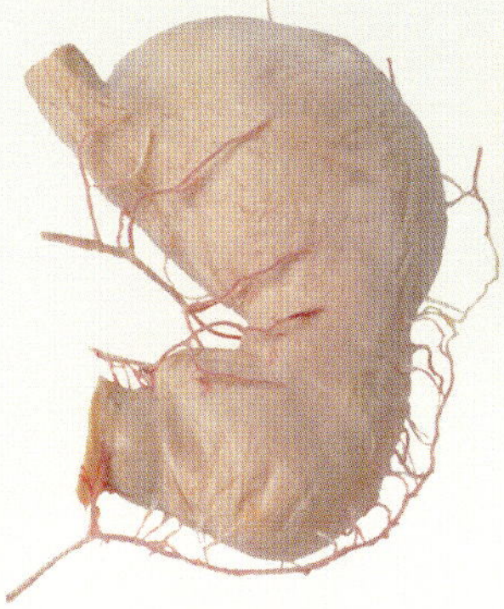

STOMACH
The stomach is the largest portion of the digestive canal. It is linked to the esophagus and the small intestine. Here, the food we eat is broken down so that its nutrients may be absorbed in our small intestine.

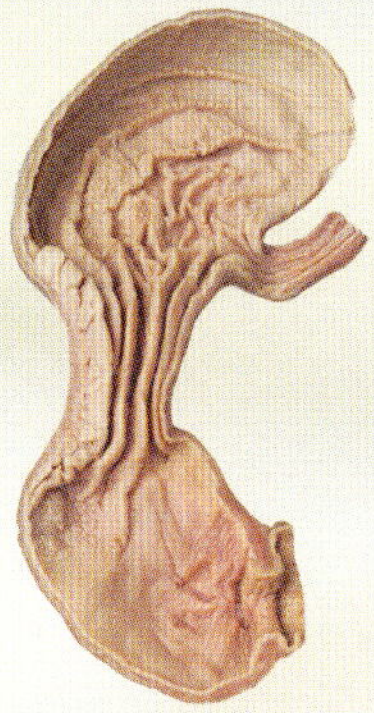

GASTRIC MUCOSA
The ridged gastric mucosa within the stomach secrete the digestive enzyme hydrochloric acid which acts to break down the foods we eat. It also secretes mucus to protect the stomach from being digested by its own processes.

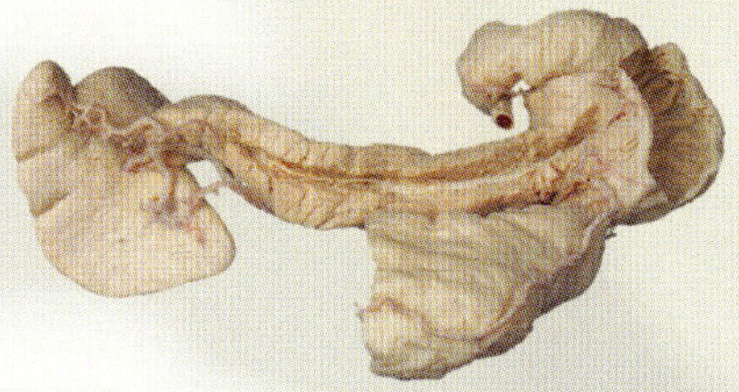

THE PANCREAS AND DUODENUM
The pancreas secretes pancreatic juice into the duodenum through the pancreatic duct. This juice contains many digestive enzymes. The pancreas also secretes insulin, which regulates the level of blood sugar and is important in the breakdown of carbohydrates. If the insulin produced is insufficient, then blood sugar rises. This disease, known as diabetes mellitus, is accompanied by the excretion of large amounts of sugar by the kidneys into the urine.

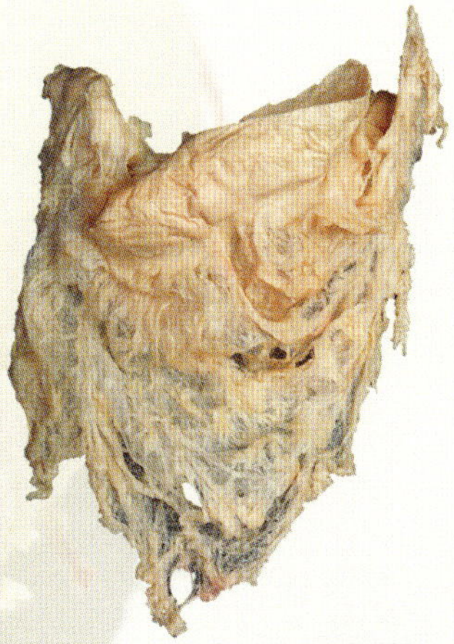

GREATER OMENTUM
The greater omentum is an apron-like fold of peritoneum that hangs down from the stomach. It supports the abdominal organs and contains vessels for blood and lymph. It also is a site for the deposit of fats and helps to support immune function. Its name derives from a Greek word for "to float", because it appears to float on the surface of the intestines.

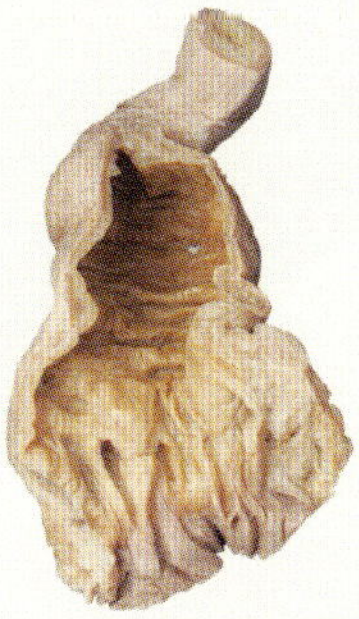

RECTUM AND ANAL CANAL
The anal canal is the end of the digestive tract. Feces is compacted and deposited here until it is excreted through the rectum.

RHYTHM

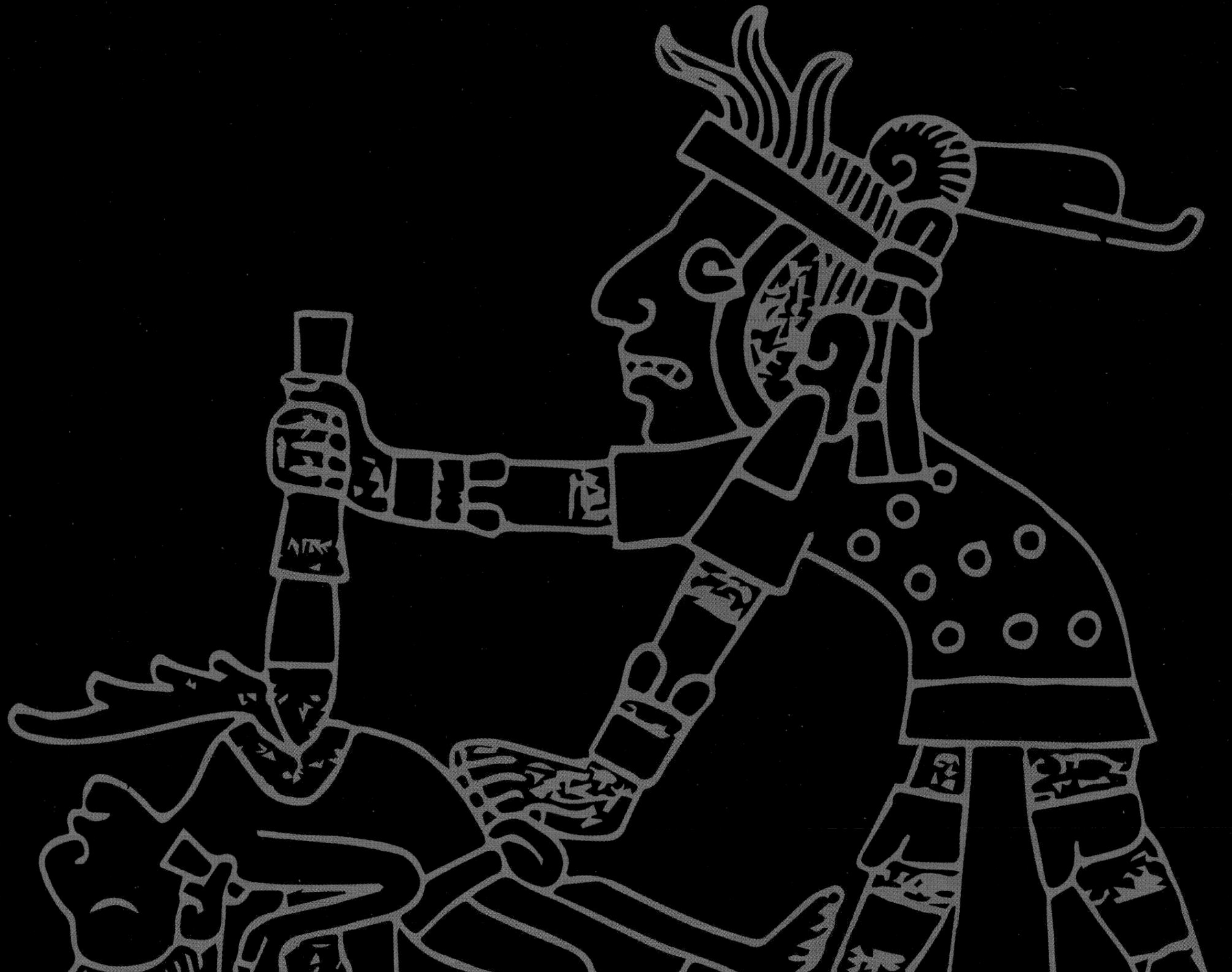

ILLUSTRATION OF HUMAN
SACRIFICE FROM AZTEC CODEX

THE CIRCULATORY SYSTEM

Like a fine lace, the circulatory system's delicate matrix weaves its way through the entire body delivering oxygen and essential nutrients to each cell in order to maintain health and vitality. Close to 100,000 miles of arteries, capillaries, and veins run through us coursing with our lifeblood and removing cellular waste. This magnificent cycle all flows according to the rhythmic drum of that all important muscle—the heart.

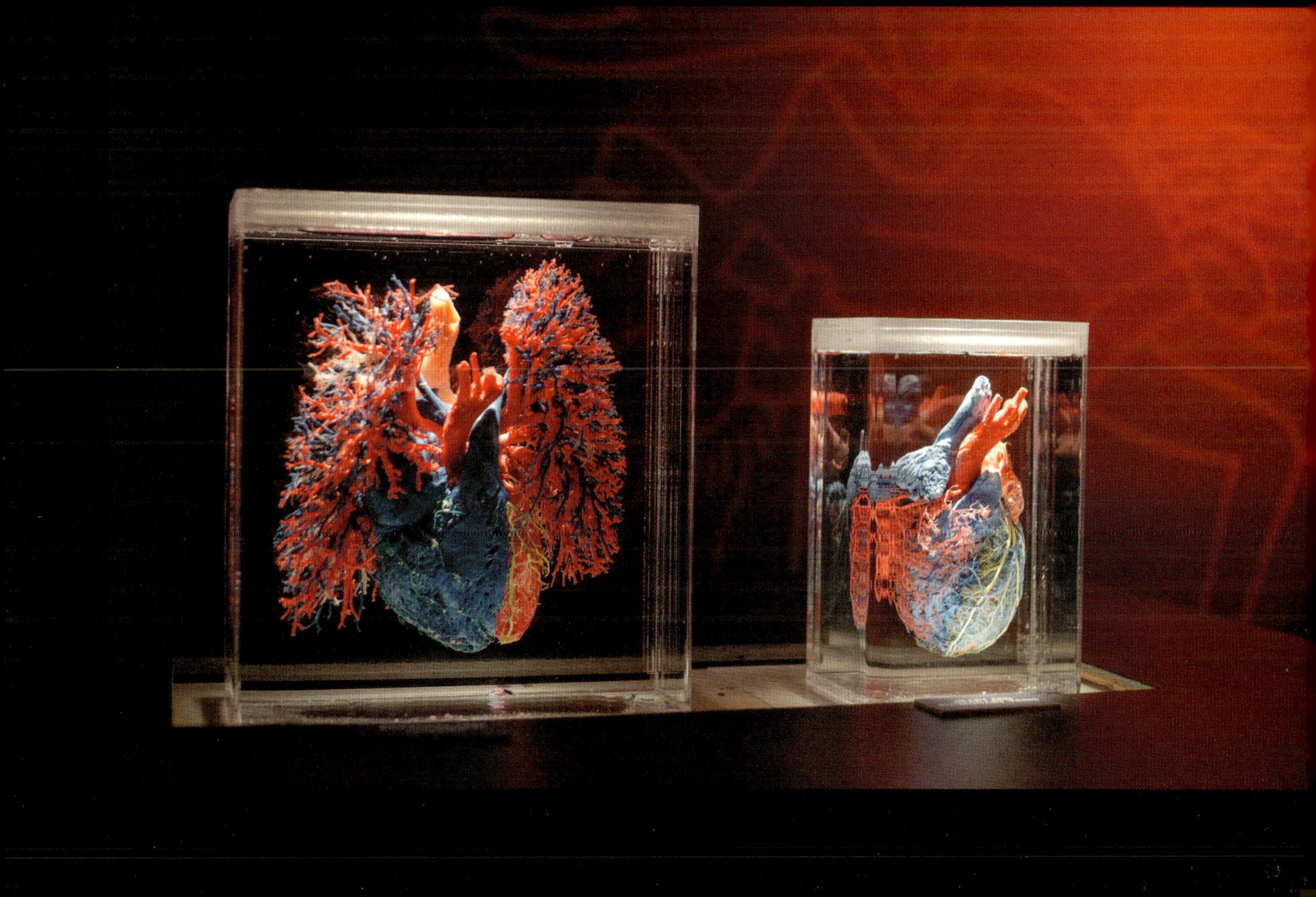

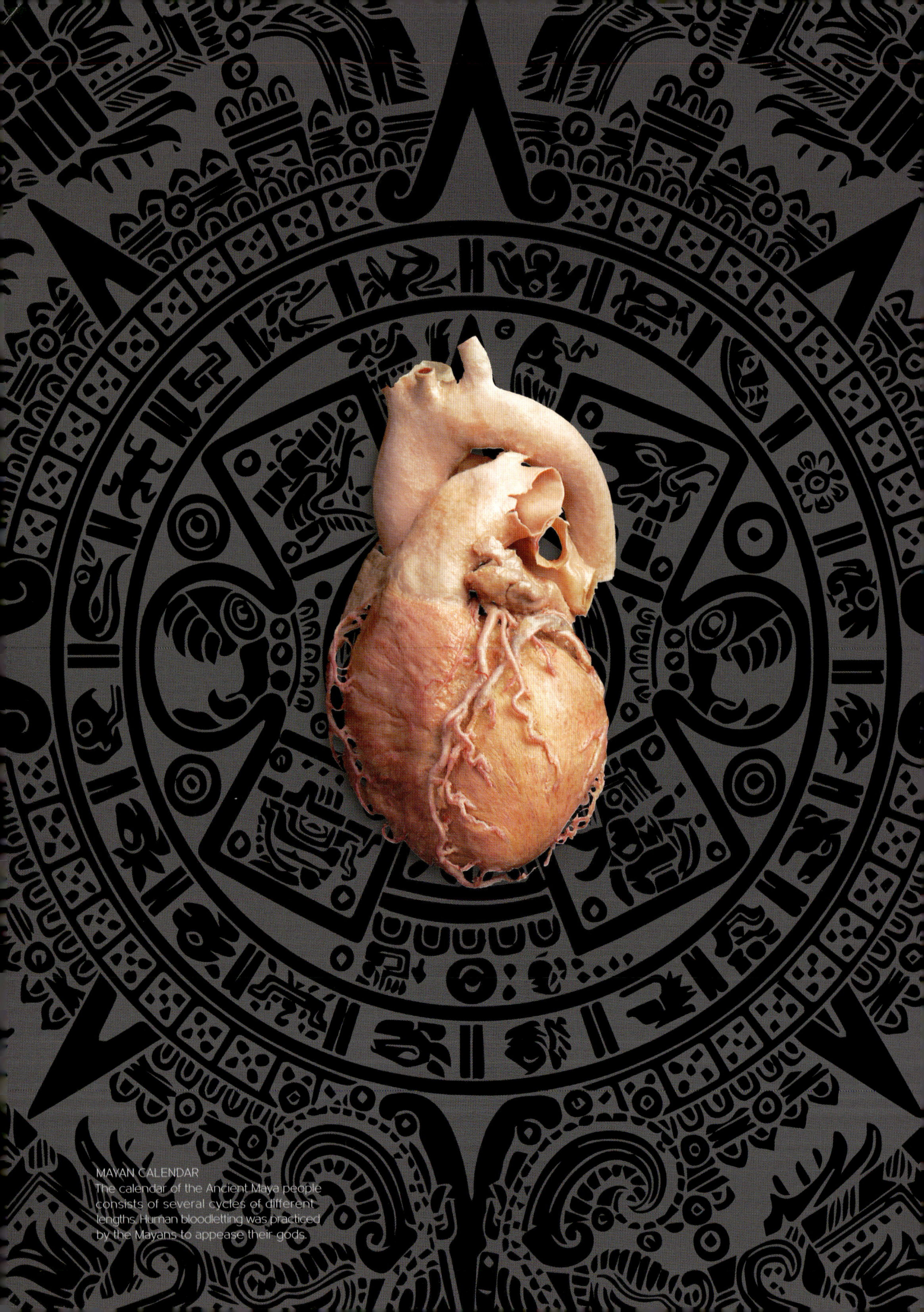

MAYAN CALENDAR
The calendar of the Ancient Maya people consists of several cycles of different lengths. Human bloodletting was practiced by the Mayans to appease their gods.

TO RHYTHM

Galloping hooves, phases of the moon, speeding train, lapping waves, sunrise, sunset...

Rhythm is a heartbeat at first. Our mother's commingling with our own.

Rhythm is breath and step. It defines the day, the seasons, the year. We create and honour the rhythms of celebration: mealtime, TGIF, harvests, birth, marriage, death.

Rhythm is community, music and language.
It commits us.

It is impossible to resist.

Repetition, structure,

Pattern, tempo, echo,

Mathematical, ecological, teleological,

Universal and organic.

Feel its pulse, its purpose–

The perfection of existence.

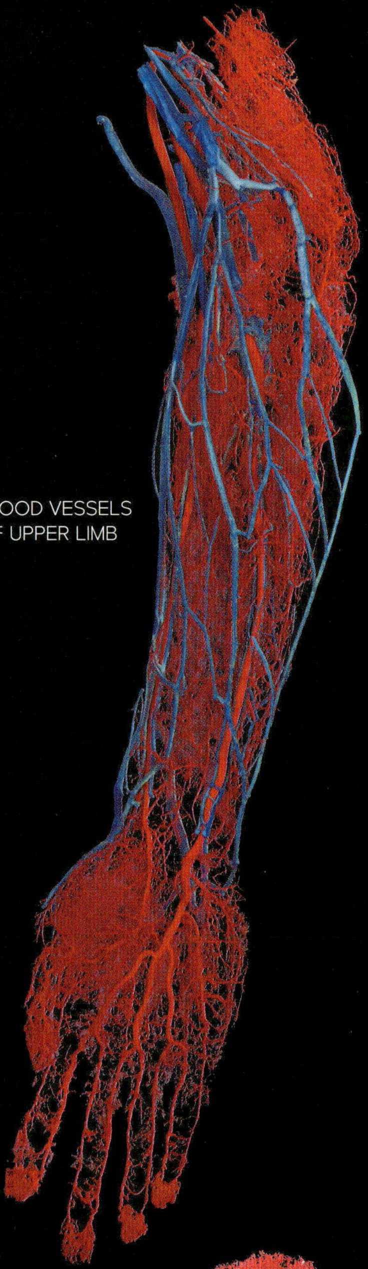

BLOOD VESSELS
OF UPPER LIMB

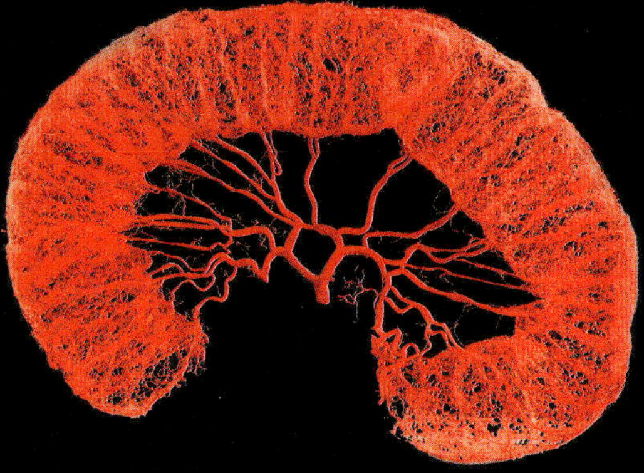

BLOOD VESSELS OF JEJUNEM

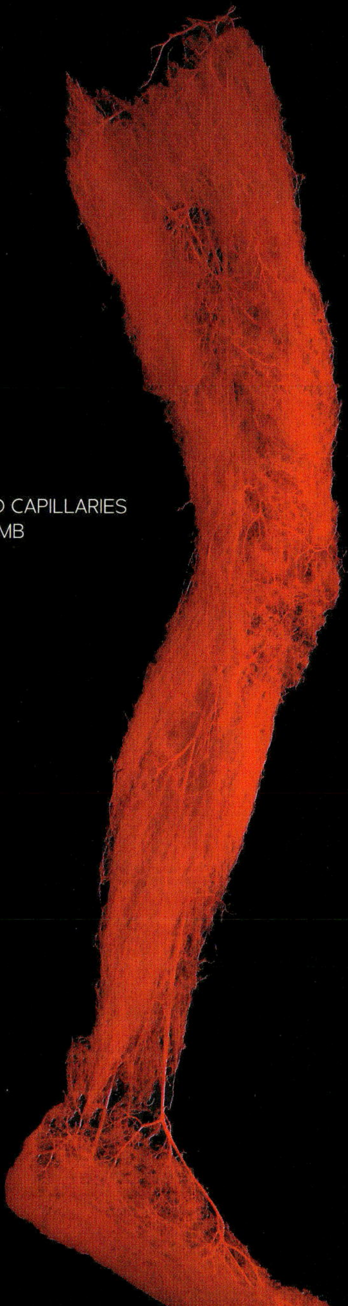

ARTERIES AND CAPILLARIES
OF LOWER LIMB

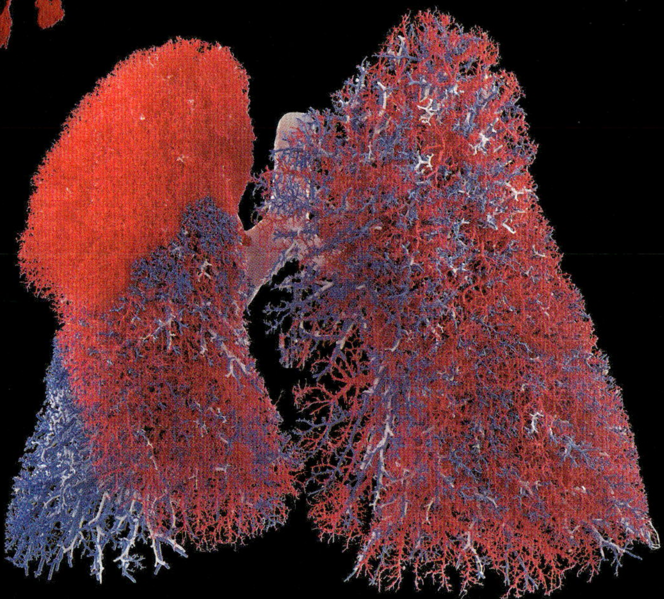

BLOOD VESSELS OF LUNG
WITH BRONCHIAL TREE

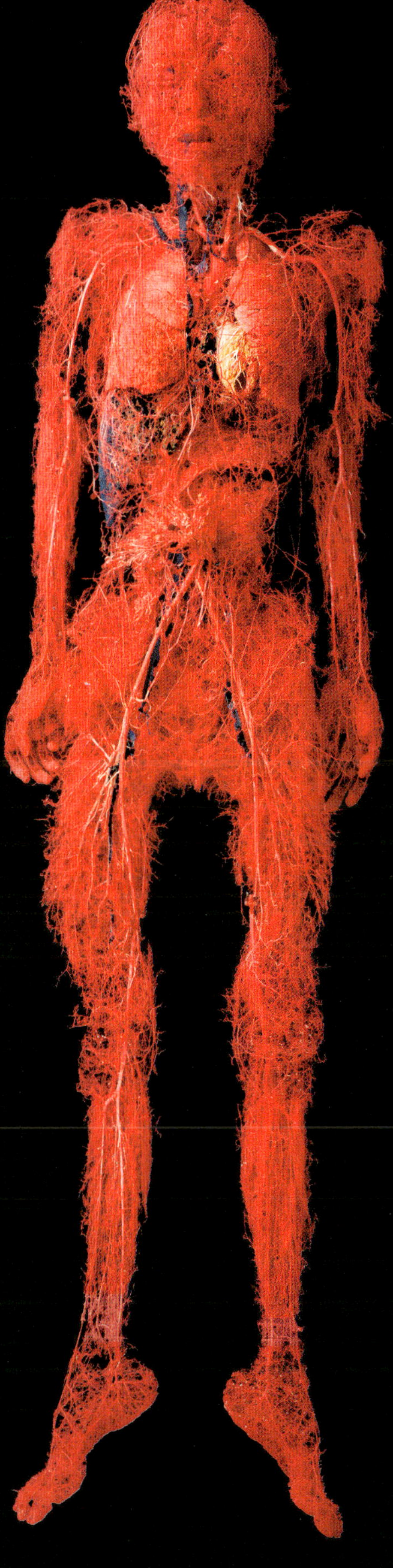

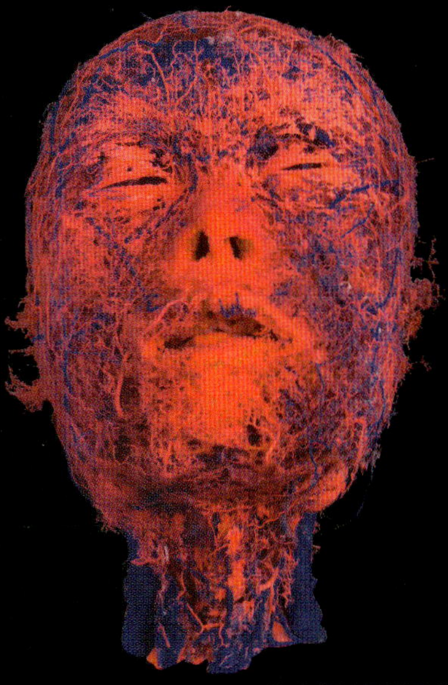

BLOOD VESSELS OF
THE HEAD AND NECK

Blood vessel castings are achieved by injecting a liquid polymer into a specimen's arteries and/or veins. Because the circulatory system is a closed system, the polymer remains in the blood vessels where it cures into a hardened rubberized material. The remaining organic tissues are removed, thus revealing the meticulous and amazing mesh of blood vessels within the body. Red indicates arteries and blue indicates veins.

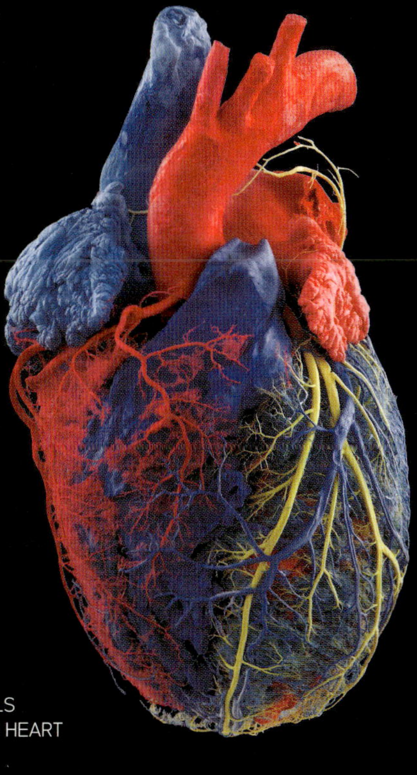

BLOOD VESSELS
SUPPLYING THE HEART

MOVE

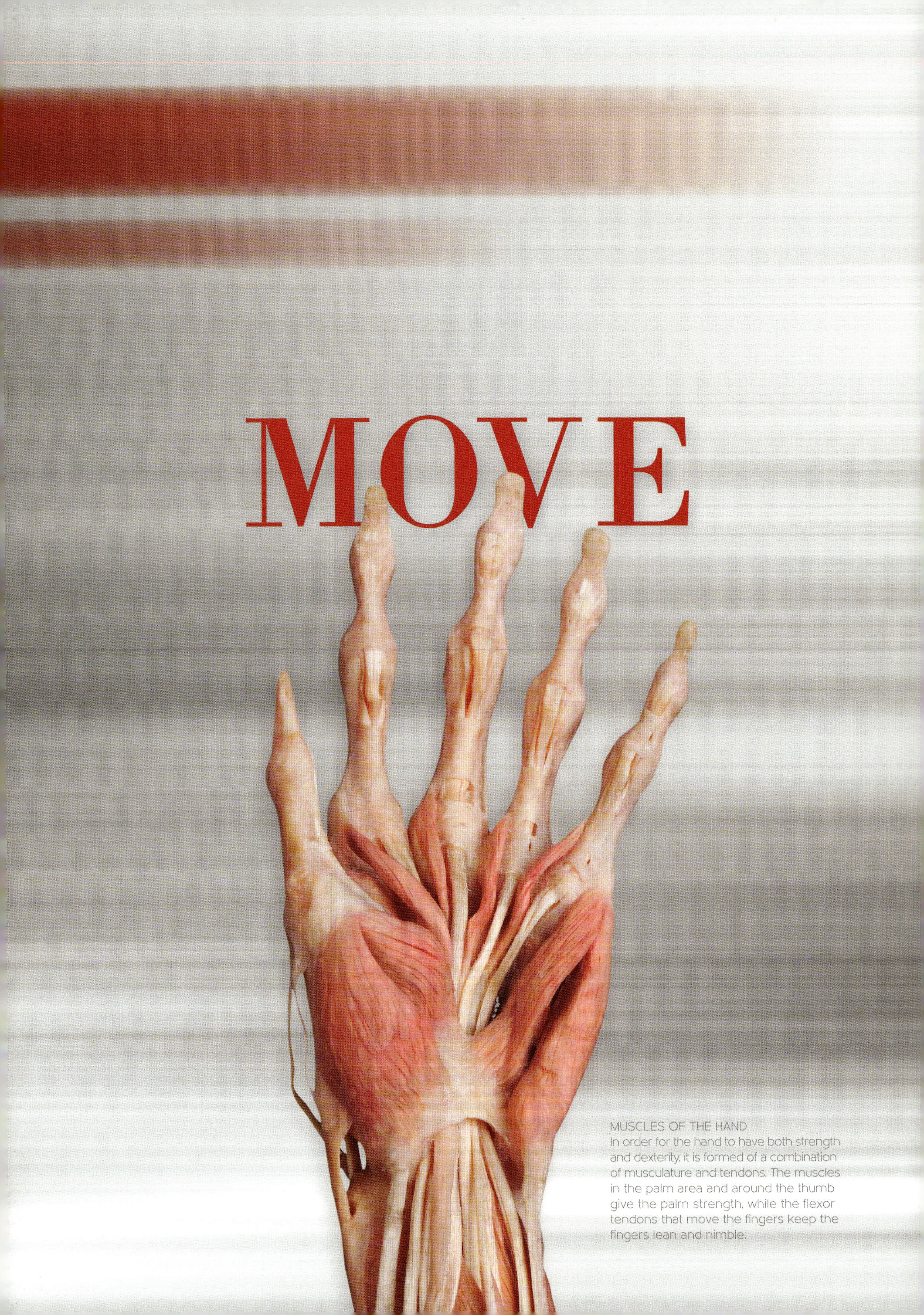

MUSCLES OF THE HAND
In order for the hand to have both strength and dexterity, it is formed of a combination of musculature and tendons. The muscles in the palm area and around the thumb give the palm strength, while the flexor tendons that move the fingers keep the fingers lean and nimble.

THE MUSCULAR SYSTEM

The nearly 840 muscles in our body move us in a variety of ways. Contractions of voluntary skeletal muscles allow us to move consciously, and generate heat to keep us warm. Involuntary muscles help move blood, food, and fluids through the body, and cardiac muscle ensures that our hearts are always beating. The human body is made up of as much as 54 percent muscle in men and 39 percent muscle in women.

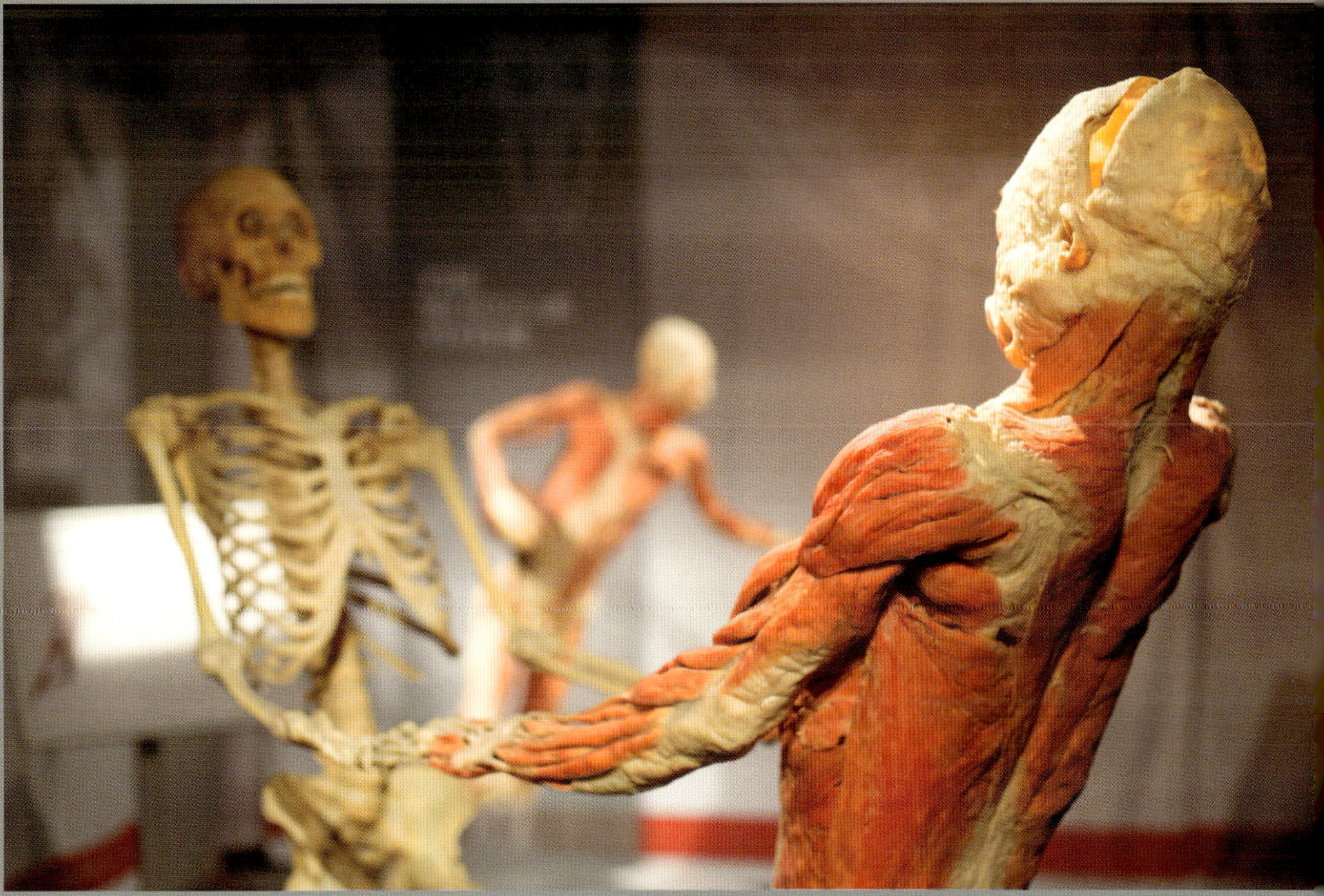

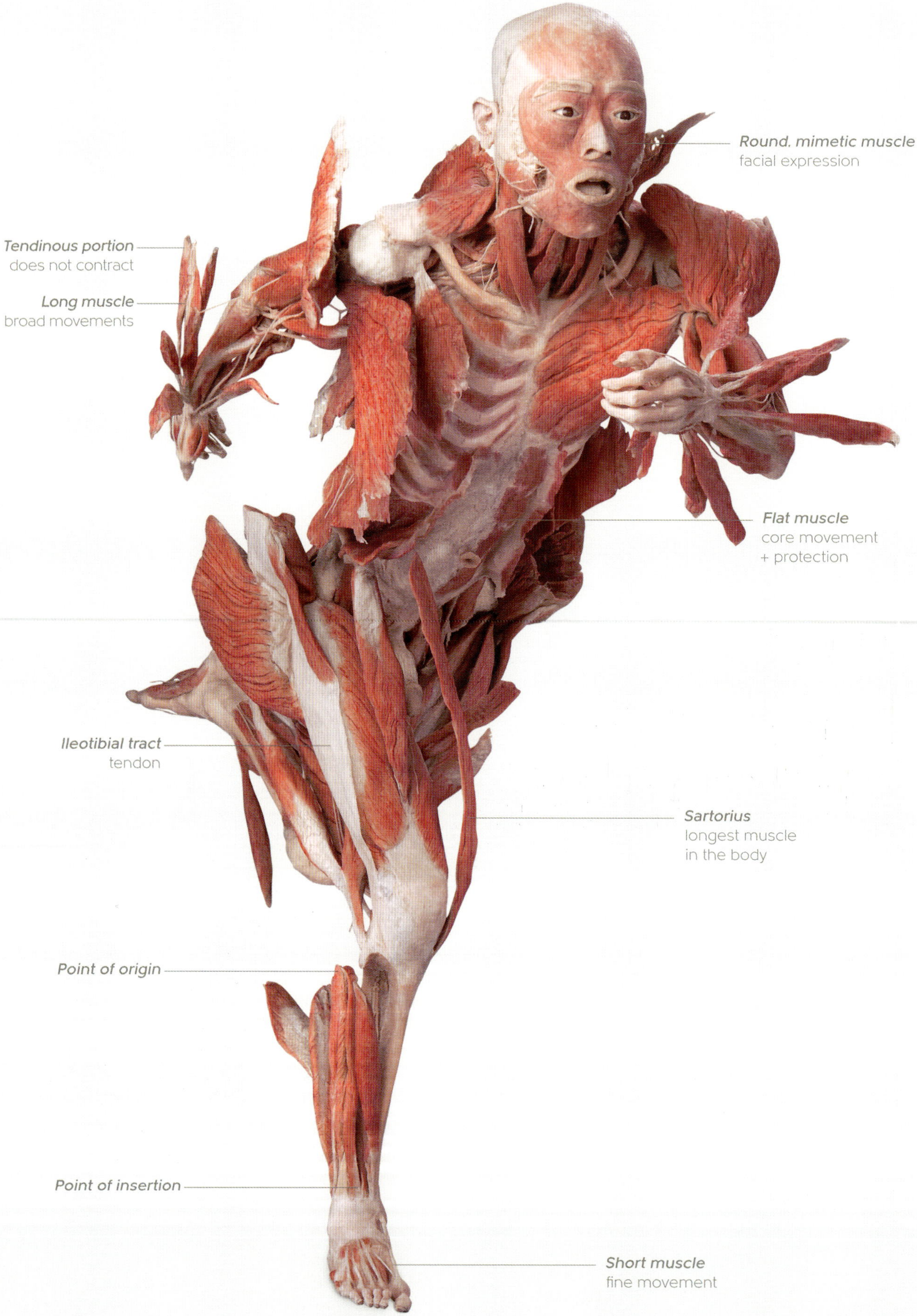

Round, mimetic muscle
facial expression
Tendinous portion
does not contract
Long muscle
broad movements
Flat muscle
core movement
+ protection
Ileotibial tract
tendon
Sartorius
longest muscle
in the body
Point of origin
Point of insertion
Short muscle
fine movement

MUSCLE ATTACHMENTS AND TYPES

Each skeletal muscle is composed of muscle and tendon. The muscular portion, also called the belly, is made up of many muscle fibers that have the ability to contract and relax. The tendinous portion, known as tendons, consists of many collagen fibers that do not contract or relax. These tendons anchor the muscle to bone, cartilage or an articular capsule. The sites of these connections are called attachments with each end of the attachment having a specific name: the point of origin indicates where muscles originate and is closest to the midline of the body; the point of insertion indicates where the muscle terminates. This specimen displays numerous muscles separated at their points of origin.

Muscles vary extremely in their shape. They may be long, short, broad or circular. Long muscles are mainly located in the extremities allowing them to contract in order to produce long-range movements. Short muscles are located in the hands, feet and head allowing for smaller, more precise motions. The flat muscles in the thorax, abdomen and back assist with movement of the core and provide protection. Circular band-like muscles are mainly found around the eyes and mouth; also known as mimetic muscle, they produce facial expressions.

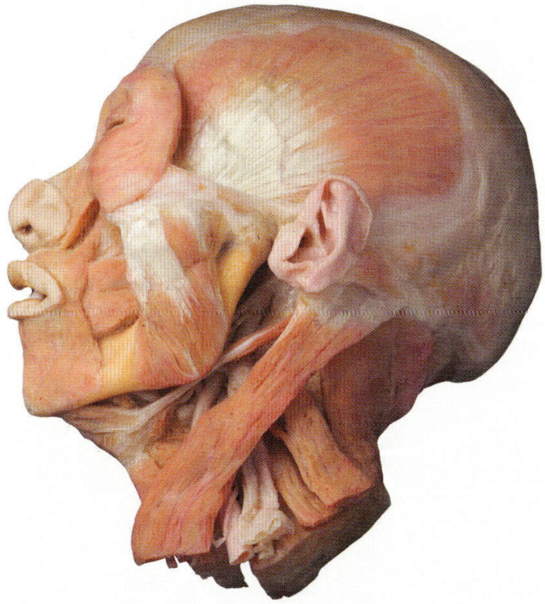

MUSCLES OF FACIAL EXPRESSION
Various facial expressions are important ways to communicate and express feelings in both humans and primates. The muscles controlling these expressions lie in the superficial layer of the face and surround the mouth, nose, eyes and ears. These muscles connect with the facial skin; even a slight contraction can produce different kinds of expressions.

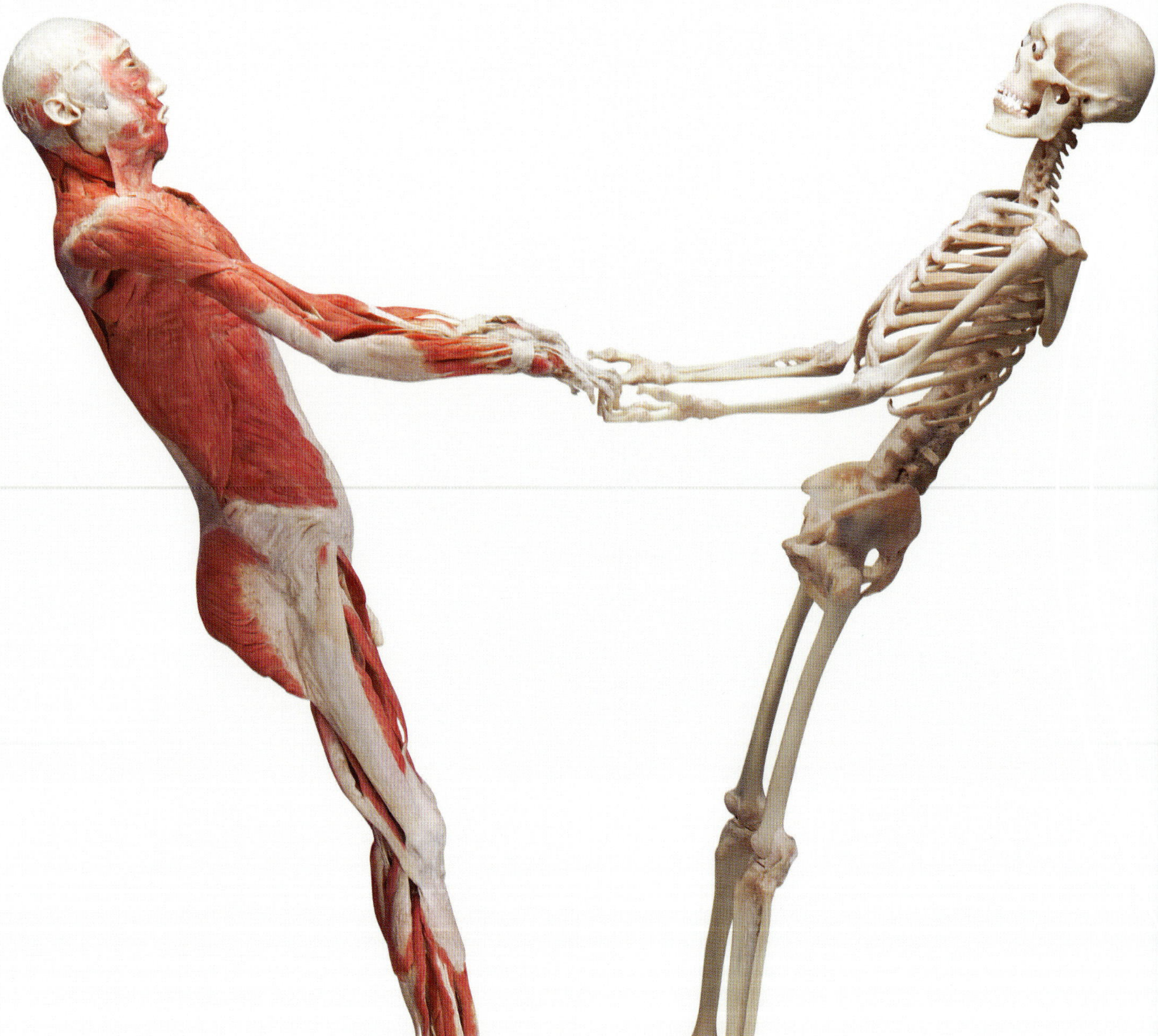

SKELETON + MUSCLES

This specimen comes from one body. The skeletal muscles were removed to demonstrate how they cover the skeleton like a perfect skin. The pose of this specimen is also a graphic example of how the skeleton and skeletal muscles rely on one another for support and movement.

The bones, joints, and muscles in the body combine to form the locomotive system. The bones act as levers that are moved (articulated) at the joints by the contraction and release of the skeletal muscles. These three groups coordinate to produce the complex movements that the body undertakes each moment. From sitting to standing to walking to running, these two systems are in constant coordination and make up more than 60% of body weight. Nearly every inch of bone in the body is covered by, and relies upon, skeletal muscle. There are over 600 skeletal muscles in the body, and they are surprisingly efficient. During contraction, only a small amount of energy is lost as heat while the rest of this energy is converted into motion.

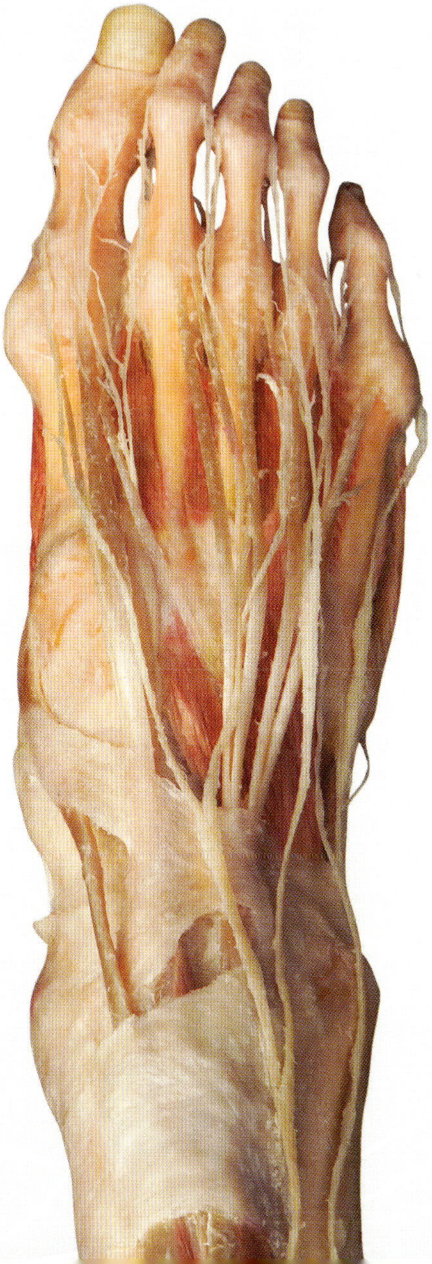

MUSCLES OF THE TOP OF THE FOOT
Assist with extending the toes, supporting the arches and maintaining balance.

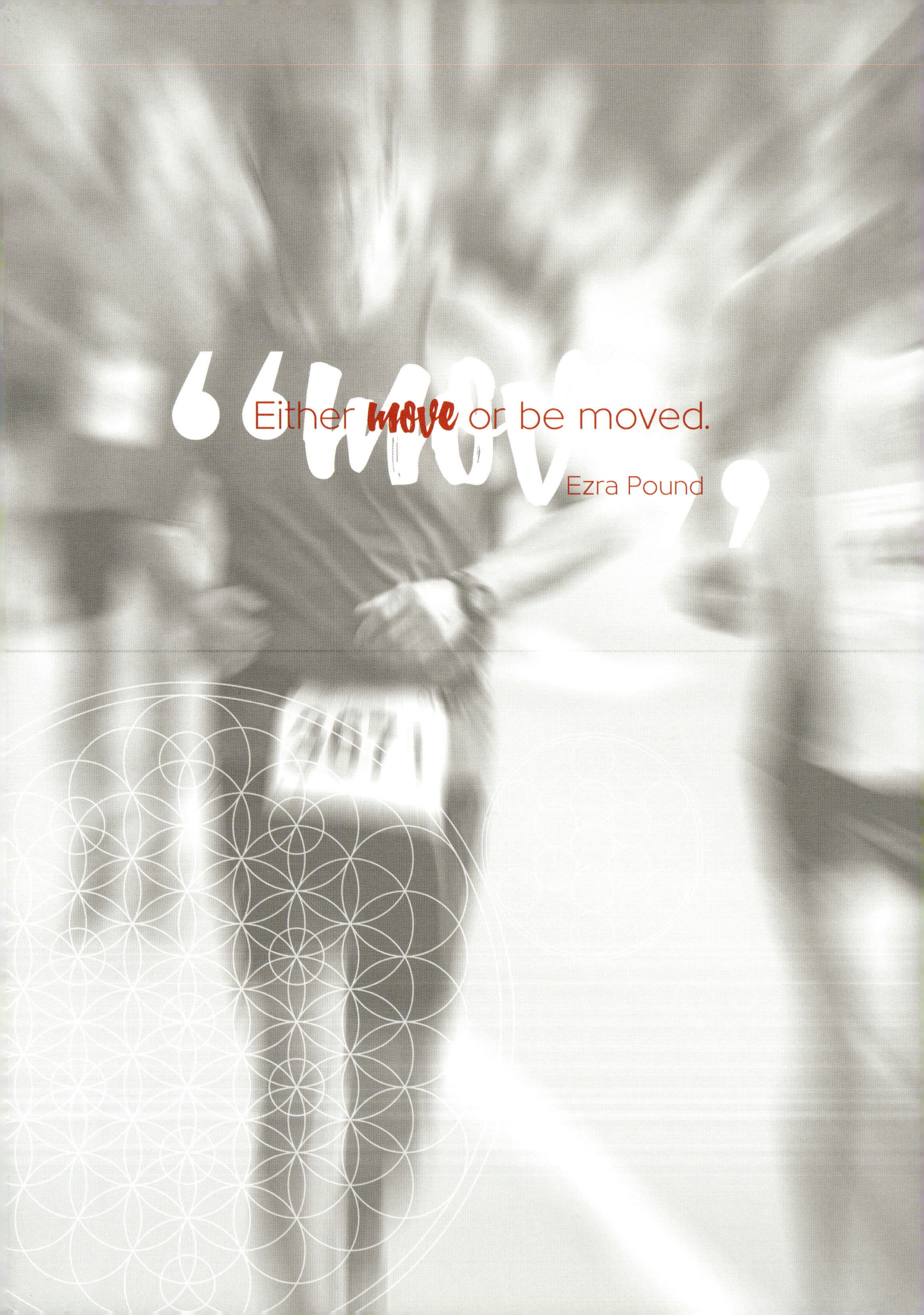
Either move or be moved.
Ezra Pound

BE STRONG

While powerful musculature has always been a component of human strength, many ancient cultures did not use human muscle as a symbol of strength. Instead, a diverse assortment of animal symbols have indicated strength across time. From the massive Bear for the knights of the Middle Ages to the small Scarab beetle in Egypt, strength has taken numerous forms in nature. It is even associated with the Oak tree in Greek and Celtic cultures.

In the same way, the word strong doesn't have a single origin; in fact, many cultures seem to have developed their own word for it. For the Italians it is closest to the word "loud," for the French, it comes from the word for fortress. In English, the word strong comes from the German "strang," a word not too far from "strange," but one that also means determined, violent, and strict.

It appears that strength in our cultural heritage has never been merely associated with muscles. It also refers to what is possible (Greek), older (German), and brave (Old English).

Then, as today, strength has always carried multiple meanings; it is one word fit for the body, mind, personality, and soul. What type of strength do you have within you?

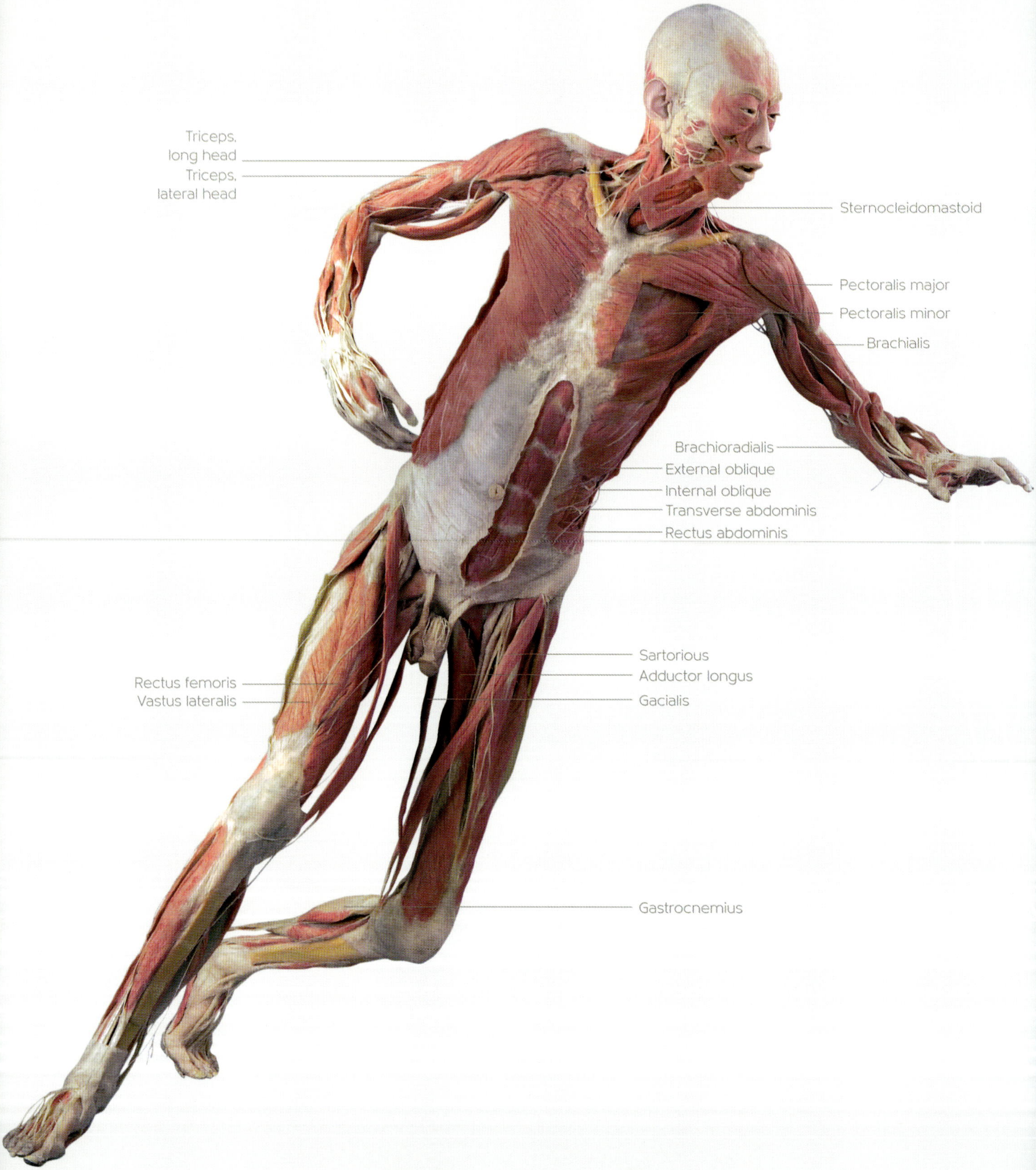
Triceps, long head
Triceps, lateral head
Sternocleidomastoid
Pectoralis major
Pectoralis minor
Brachialis
Brachioradialis
External oblique
Internal oblique
Transverse abdominis
Rectus abdominis
Sartorious
Adductor longus
Gacialis
Rectus femoris
Vastus lateralis
Gastrocnemius

MUSCLE LAYERING, PAIRING + MOTOR CONTROL

The body's skeletal muscles consist of several layers that pass from superficial (closest to the skin) to deep. This specimen shows the relationships of those different layers. In many parts of the body, there are more than two layers of muscles that overlap each other. For example, the muscles of the lateral abdominal wall can be divided into three layers. The muscular fibers of each layer are arranged along different directions, which form a strong abdominal wall. When contracting at the same time, these muscles can increase abdominal pressure to assist with breathing, lifting, running coughing, and childbirth among other activities.

Typically, each joint connects with at least two groups of muscles that oppose the action of the other. In this way, muscles can contract to move a joint while the opposing muscles relax. For example, the muscles on the underside of your forearm contract actively to flex the wrist joint downward, while those on top contract to extend it upwards. Muscles paired in this way are known as antagonists.

Fine movements are not completed by a single muscle. Two or more muscles coordinate together to create these delicate and dexterous motions. Sometimes muscles must work at a distance as well. For example, the flexors that control the fingers are located in the forearm and work by pulling tendons; muscles in the hand would make it too bulky to perform fine movements.

Those ***move*** easiest who
have learned to dance.

Alexander Pope

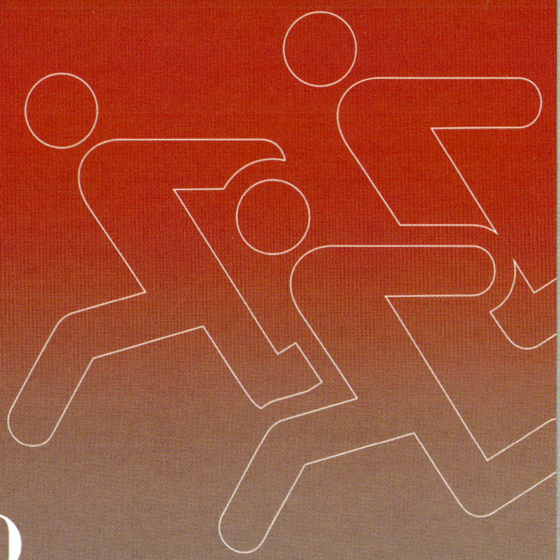

TO MOVEMENT

Perfect swing, shopping trip, runner's stride, drinker's draught, dancer's gentle footfall. With absolute certainty and precision–even in our indecision–the muscles cannot lie. Living levers and pulleys, coursing with blood and heat, they animate our bones. They can make us the finest athlete, the weekend warrior, the laborer, the gardener, the DIY enthusiast, the surgeon, the poet, the lover, the couch potato.

Facial twitch, nostril flare, open arms, clenched fist, work of art, act of war–every move we make defines us. With no effort, the muscles move us through the day and never say a word. The language of the sinews is the unspoken reality of now.

The history of the world is made with muscles. They are the beautiful strings that move the marionette. And we are the actors, puppets and puppeteers, in this little play called life.

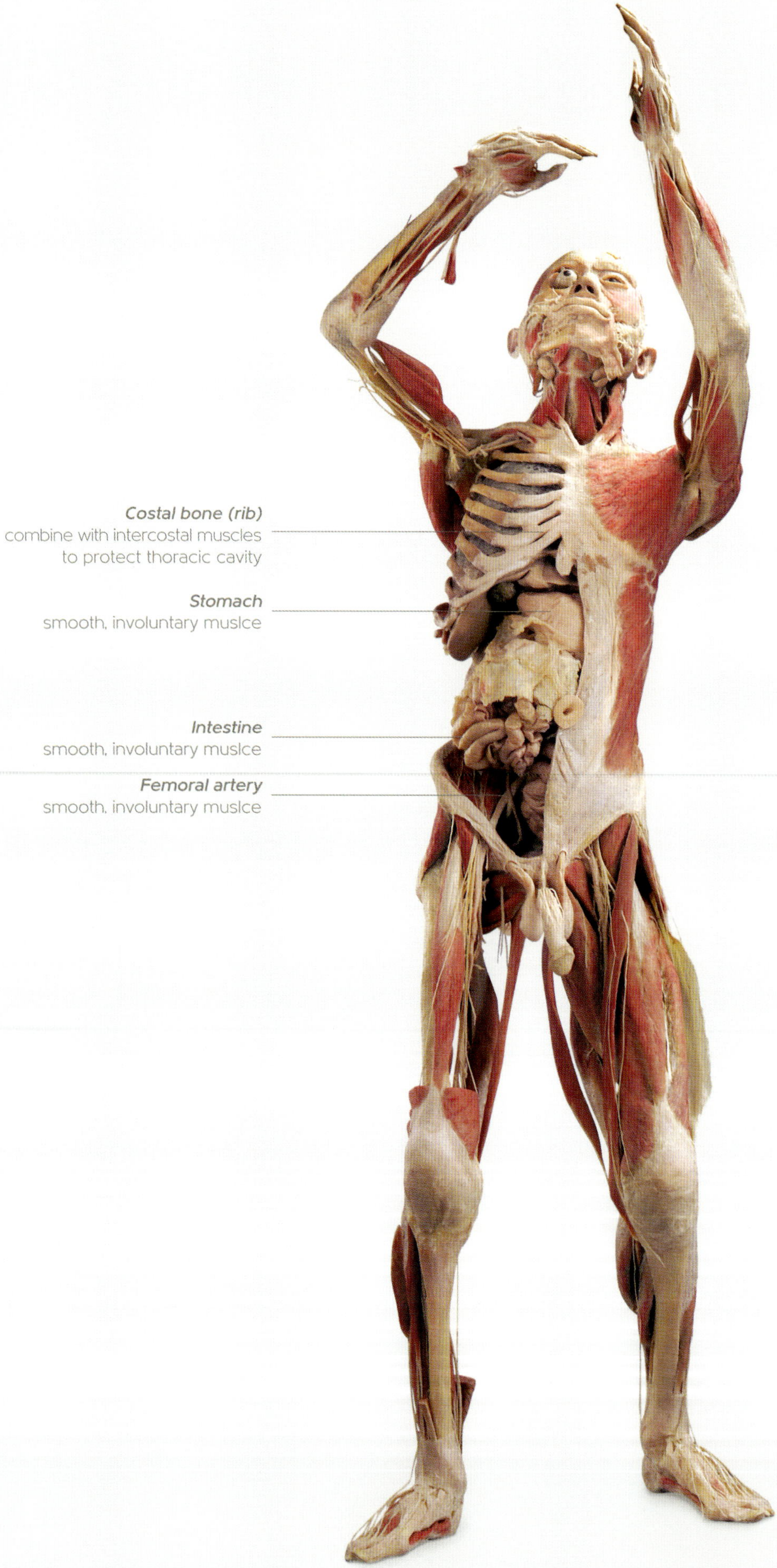
Costal bone (rib)
combine with intercostal muscles
to protect thoracic cavity
Stomach
smooth, involuntary muslce
Intestine
smooth, involuntary muslce
Femoral artery
smooth, involuntary muslce

PROTECTION + TONE

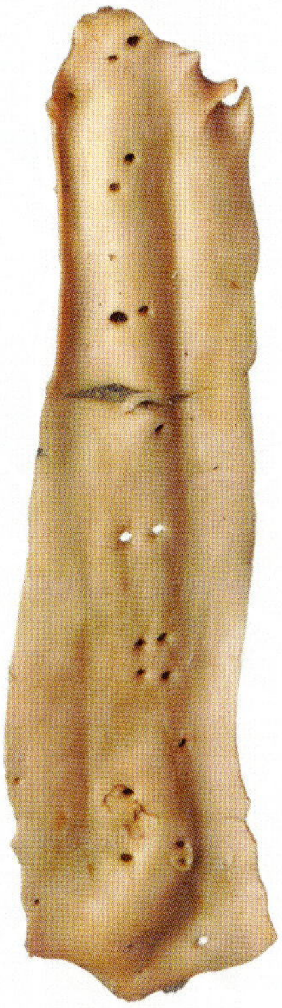

Along with making the body move, muscles also play an important role in protection.

In some locations, the skeleton and muscles work together in order to protect the inner organs. For example, the intercostal muscles between the ribs assist in forming the rib cage that protects the visceral organs; and the abdominal muscles must play the role of protector of the belly with no skeleton to assist them.

Muscles also provide a tremendous amount of support to the body, keeping it stable. This is accomplished through muscle tone, fine-tuning of the muscles at all times to keep you balanced, your lips together, your face taut and your posture correct. Even when sitting down or laying completely still, your muscles are still keeping you in check.

INVOLUNTARY MUSCLES
When we think of muscles, skeletal muscles first come to mind. This is because these are under our conscious control. But our bodies contain hundreds of muscles we do not control as well. These involuntary muscles function without our thinking about them. Smooth involuntary muscle controls the movement within organs such as the esophagus, stomach, intestines, bronchi, uterus, urethra, bladder, and blood vessels. Striated cardiac muscle controls the function of the heart.

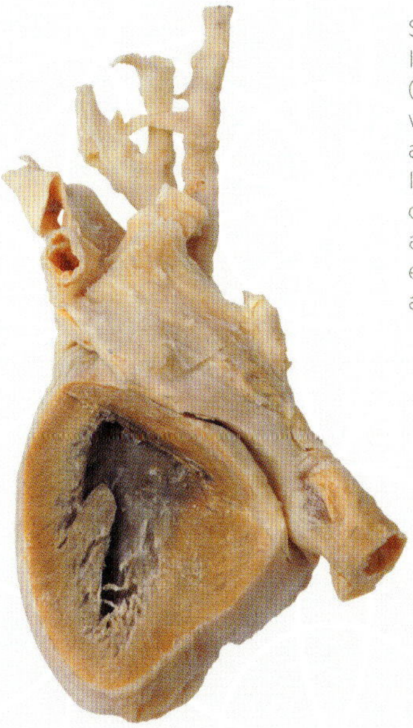

SMOOTH MUSCLE TISSUE
In digestive organs such as the stomach (below) and aorta (above), the constant work of smooth muscle keeps food and blood moving through our bodies. In the digestive tract, muscle contractions called peristalsis create an endless flow to advance nutrients. In the arteries, muscles expand and contract to deliver blood and maintain blood pressure.

CARDIAC MUSCLE
Cardiac muscle makes up most of the heart. It is the most active and strongest muscle in the body. An involuntary muscle, cardiac muscle cannot be controlled by your conscious thought.

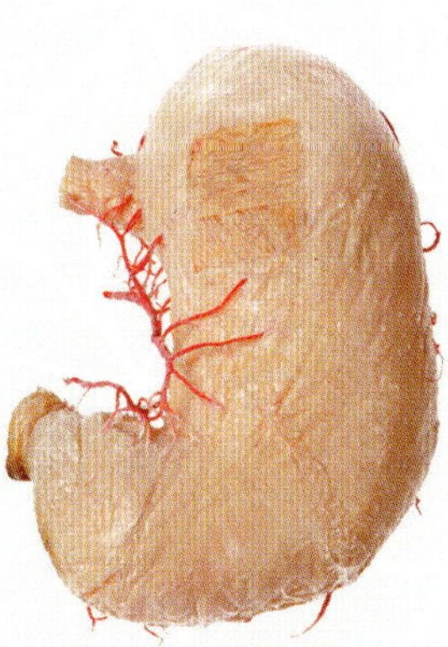

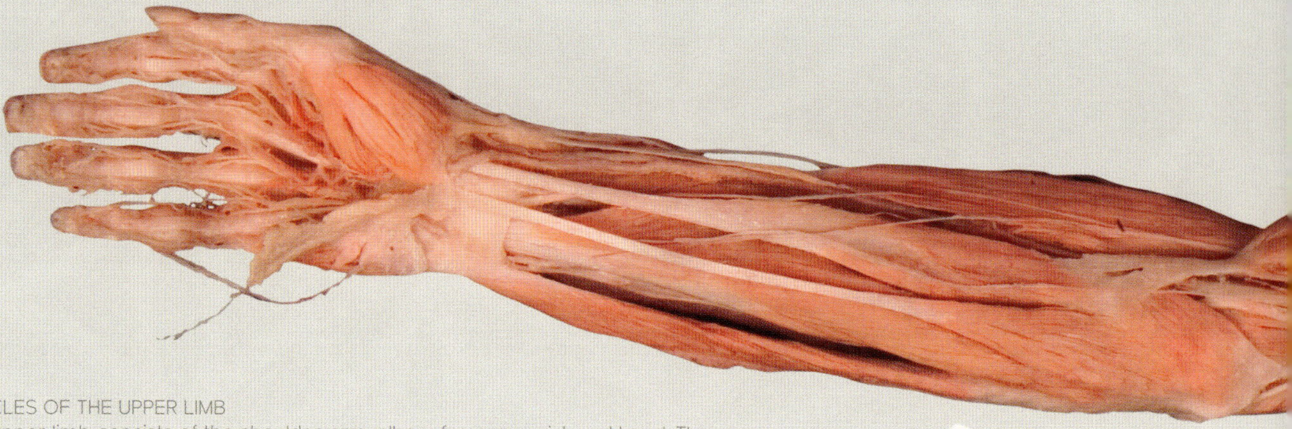

MUSCLES OF THE UPPER LIMB

The upper limb consists of the shoulder, arm, elbow, forearm, wrist, and hand. They move the joints in different directions. The deltoid moves the arm outwards, forwards, and backwards. The biceps flexes the forearm. The triceps extends the forearm. Some muscles combine to rotate the forearm, as in the act of turning a screw driver.

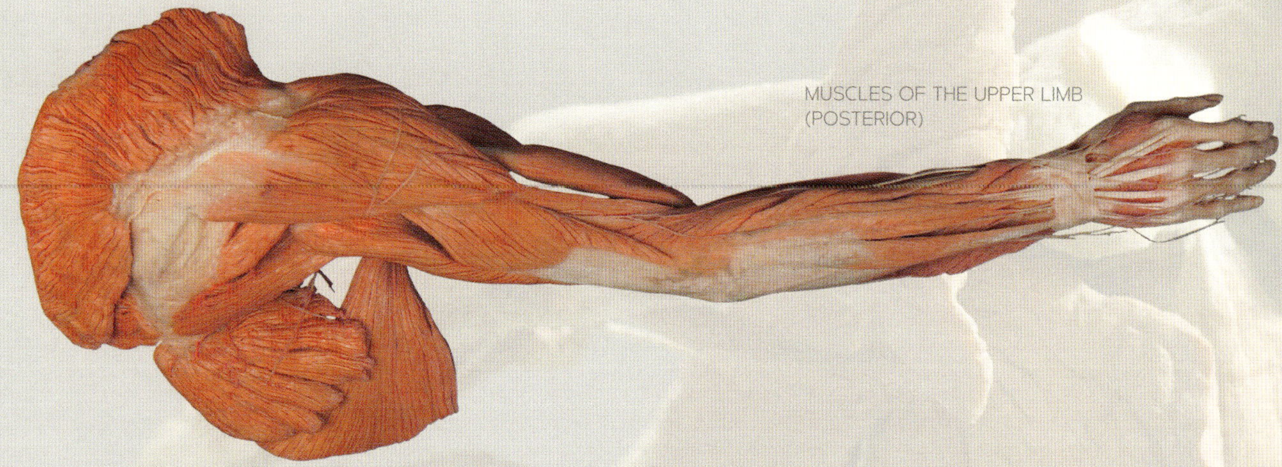

MUSCLES OF THE UPPER LIMB (POSTERIOR)

MUSCLES OF THE LOWER LIMB

The lower limb consists of the buttocks, pelvis, thigh, knee, leg, and foot. Along with their associated bones, they keep us upright and support body weight as we walk. Thus, they are stronger than those of the upper limb. The combined four muscles of the quadriceps in the thigh form the largest muscle of human body, and the sartorius is the longest.

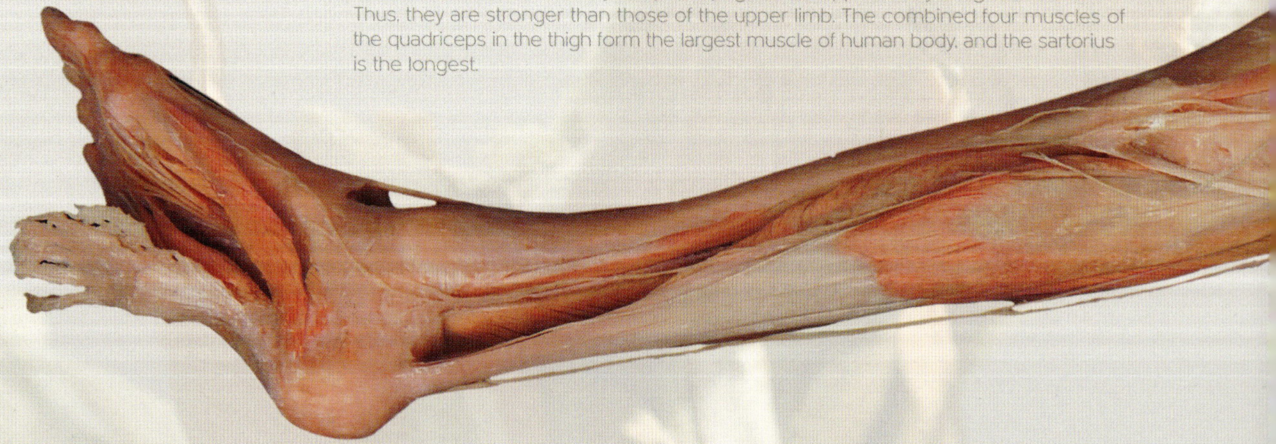

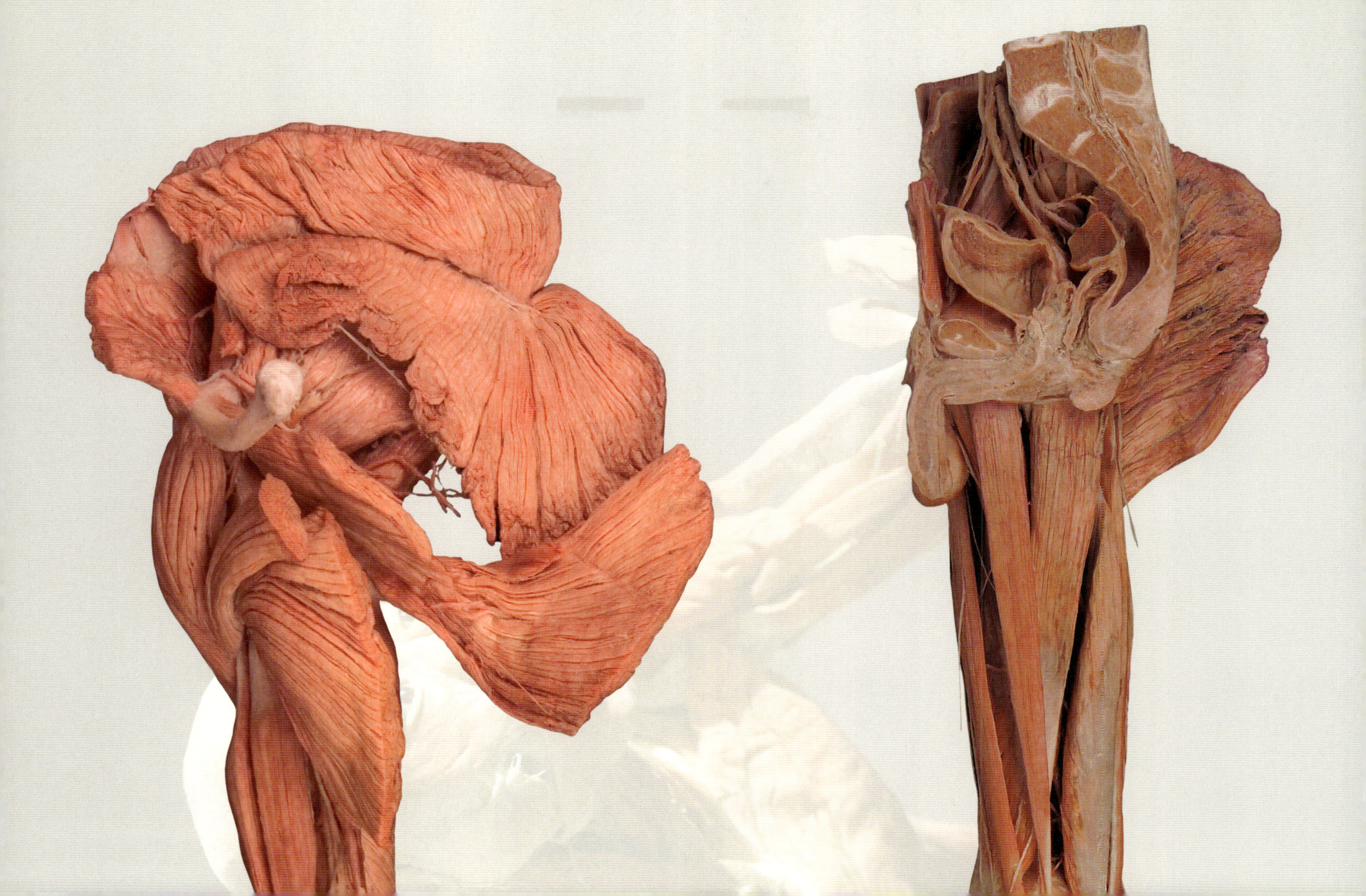

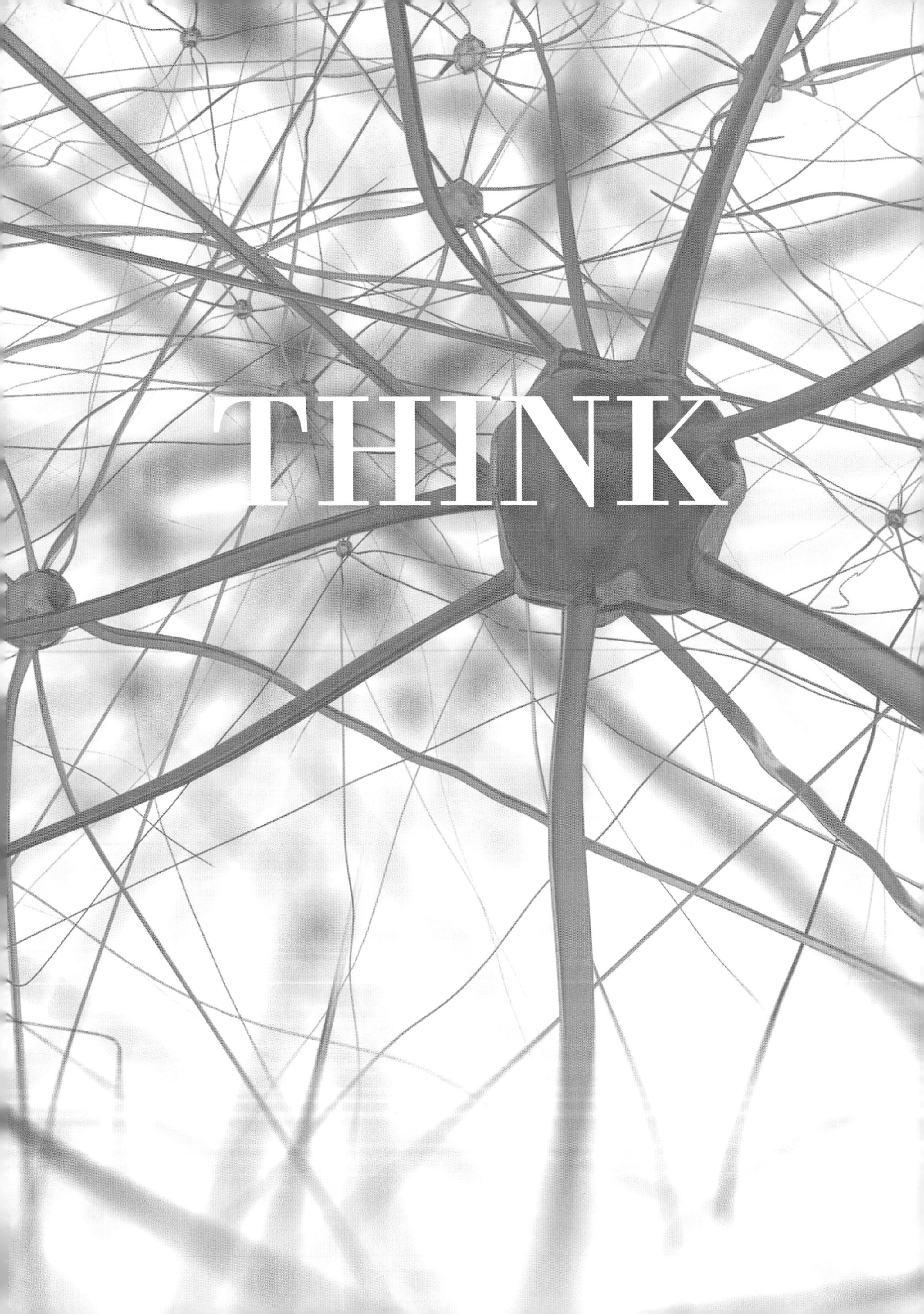
THINK

THE NERVOUS SYSTEM

The most complex computer network on the planet, the human nervous system coordinates a countless number of operations each second. Our five senses, physiological functions, thoughts, hopes and dreams are all under its continuous control. The brain consists of 100 billion 100 billion neurons and 45+ miles of nerve networks that are in constant communication in order to keep your body in a state of balance and your mind focused on the task at hand.

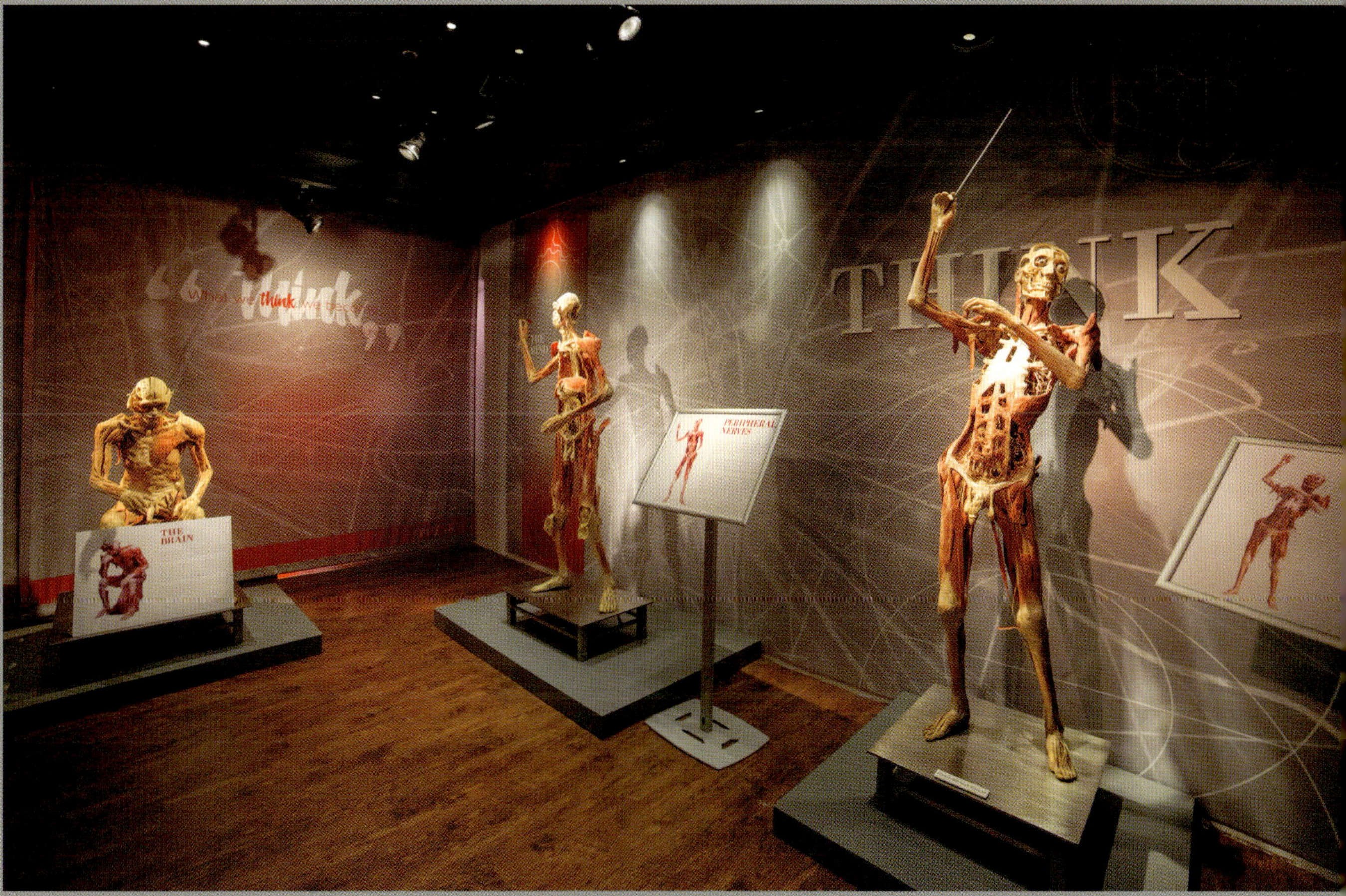

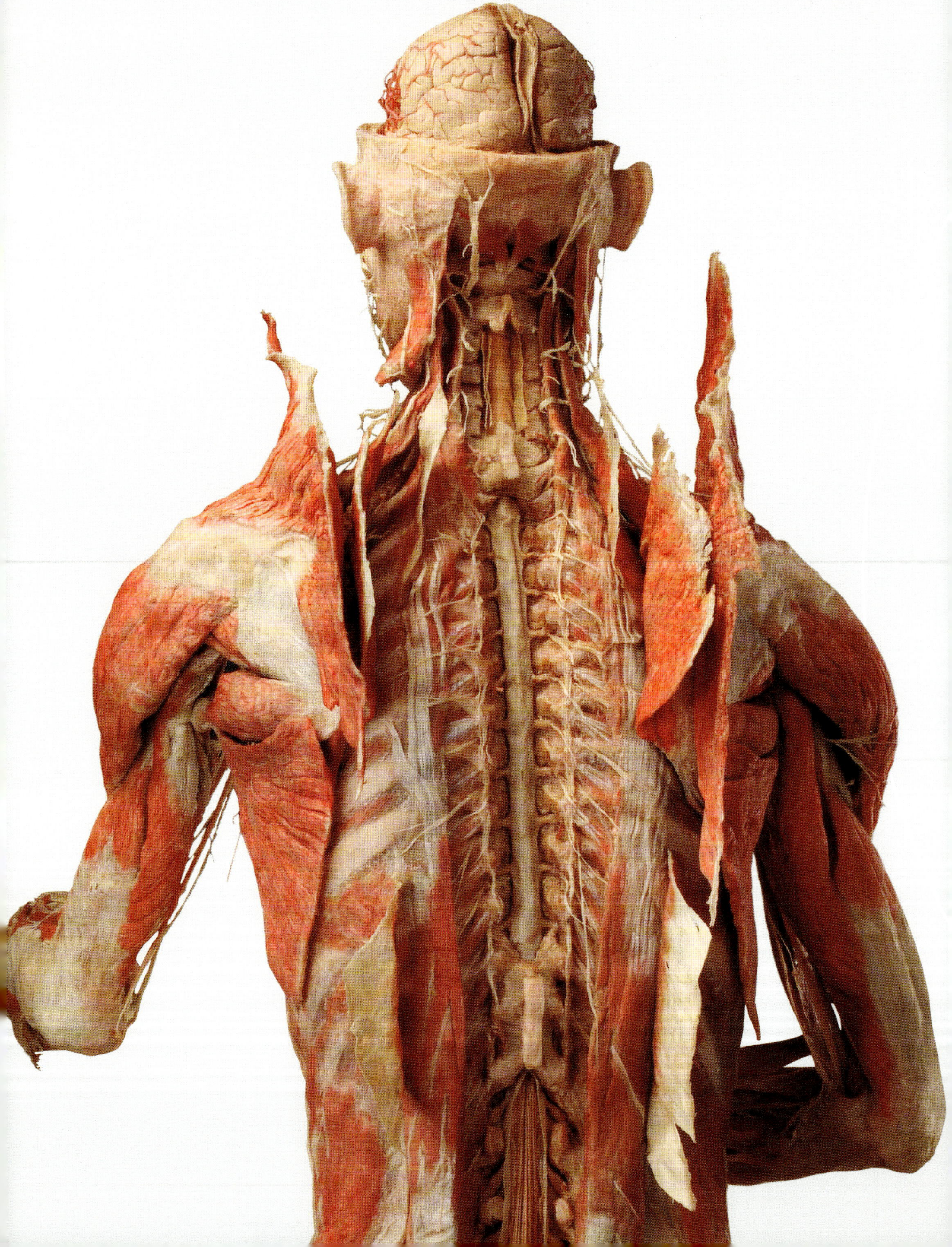

CENTRAL COMMAND

The central nervous system's two major components are the brain and spinal cord–the Messager and Messenger of the body. The brain receives messages from the rest of the body via the spinal cord and sends messages throughout the body in the same way.

Situated in the cranial cavity, the brain is well protected by the thick bones of the skull and three layers of meninges. The spinal cord, which attaches to the brain at the base of the skull, is also well protected by the vertebrae and the spinal meninges. In this dissection, both the brain and spinal cord are revealed in all their glory allowing you to see the first information superhighway known to humankind.

The spinal cord is between 17 and 18 inches long and gives rise to the spinal nerves that emerge from special openings in the vertebrae. They transmit information from the spinal nerves to the brain to produce feelings such as temperature, pressure, and discomfort.

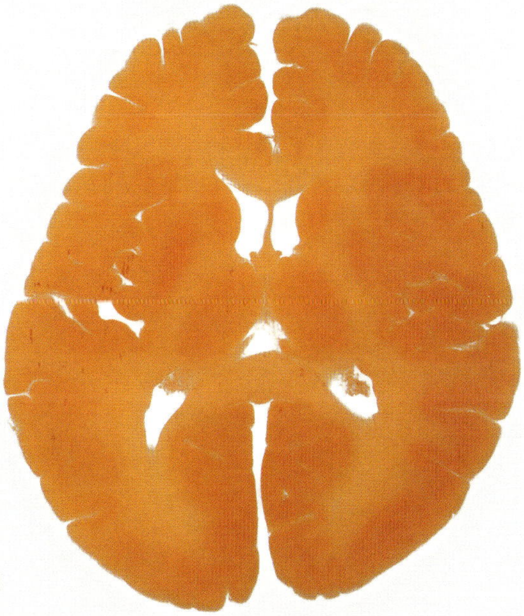

CROSS SECTION OF BRAIN

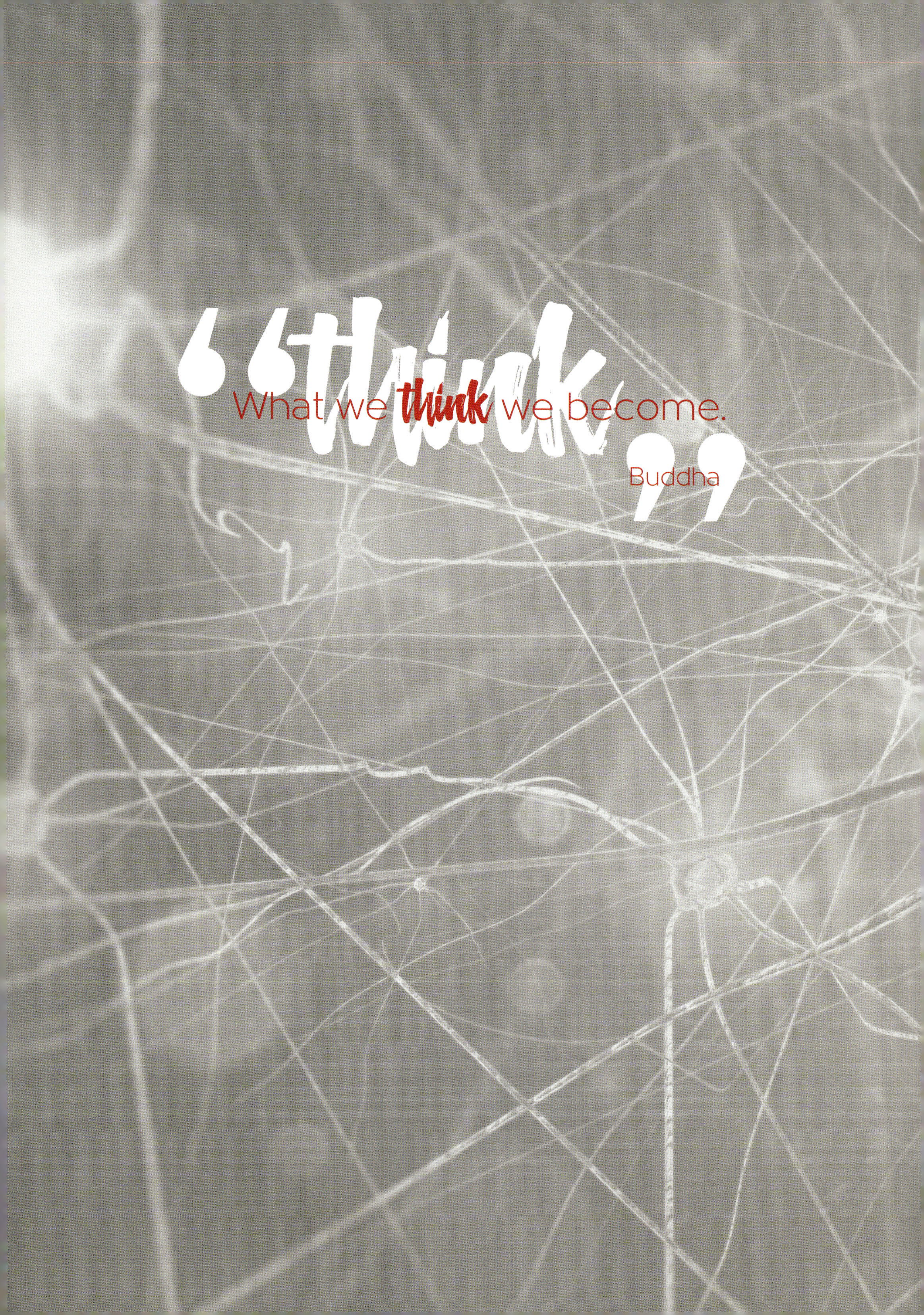
think
What we think we become.
Buddha

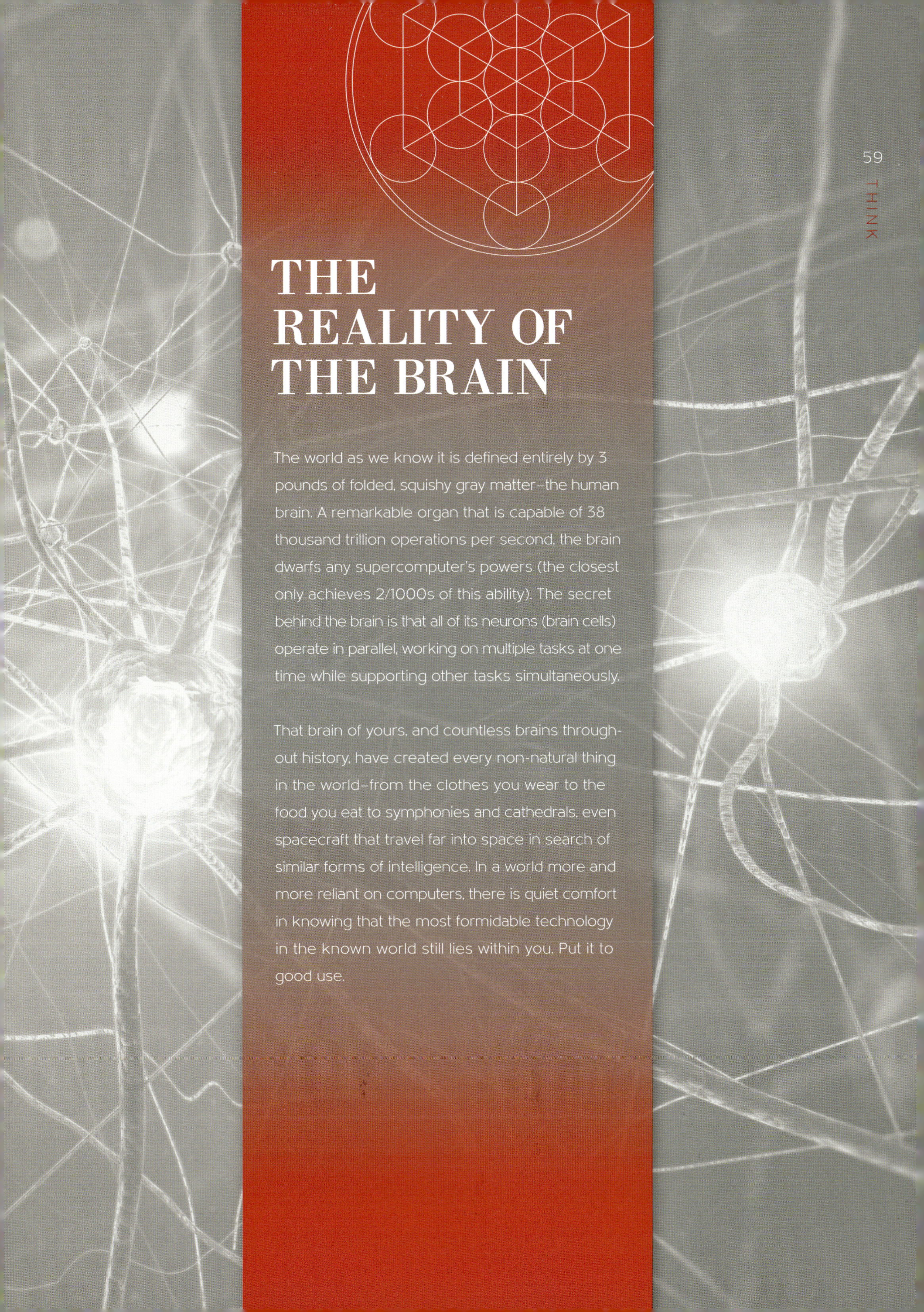

THE REALITY OF THE BRAIN

The world as we know it is defined entirely by 3 pounds of folded, squishy gray matter–the human brain. A remarkable organ that is capable of 38 thousand trillion operations per second, the brain dwarfs any supercomputer's powers (the closest only achieves 2/1000s of this ability). The secret behind the brain is that all of its neurons (brain cells) operate in parallel, working on multiple tasks at one time while supporting other tasks simultaneously.

That brain of yours, and countless brains throughout history, have created every non-natural thing in the world–from the clothes you wear to the food you eat to symphonies and cathedrals, even spacecraft that travel far into space in search of similar forms of intelligence. In a world more and more reliant on computers, there is quiet comfort in knowing that the most formidable technology in the known world still lies within you. Put it to good use.

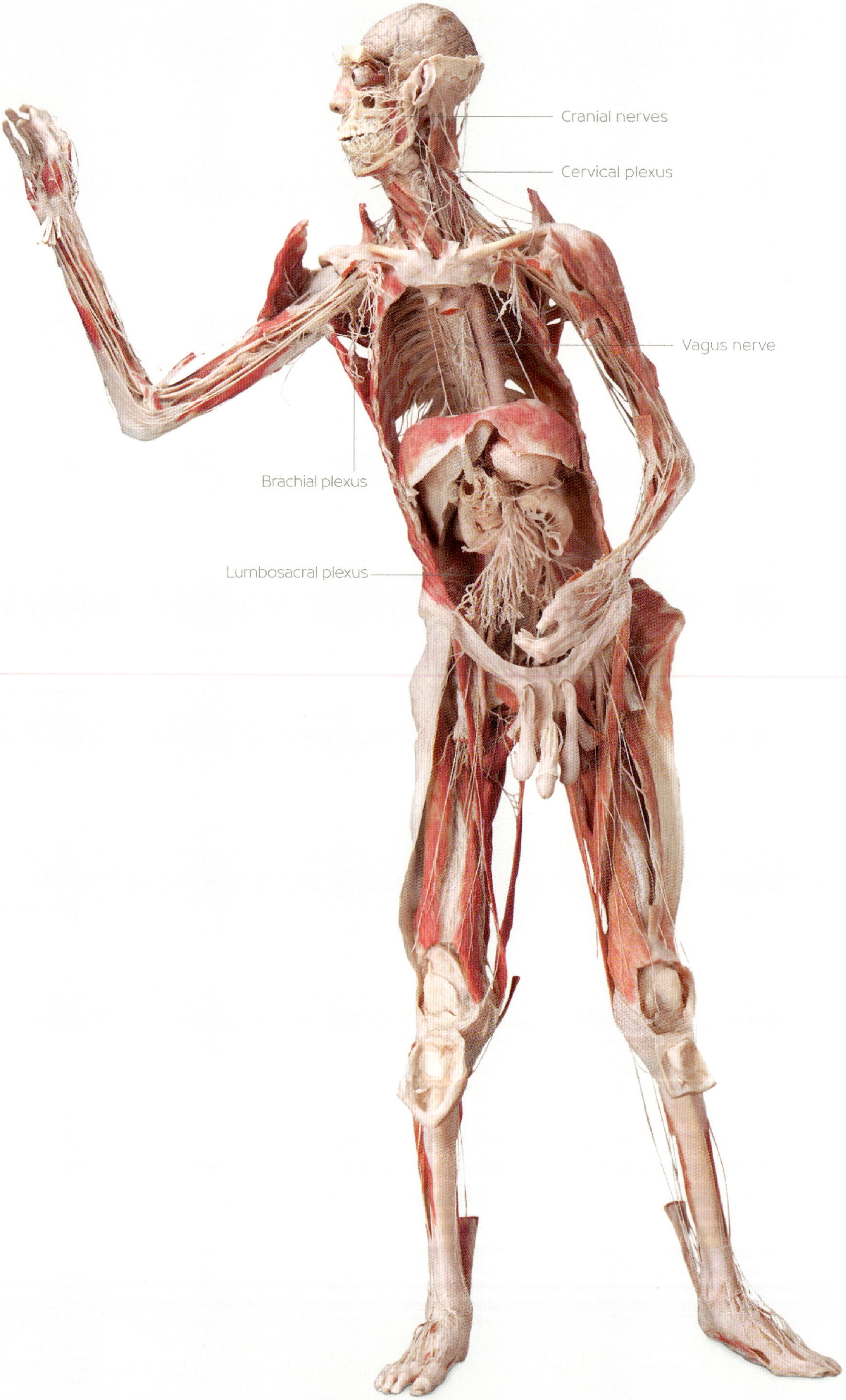
Cranial nerves
Cervical plexus
Vagus nerve
Brachial plexus
Lumbosacral plexus

PERIPHERAL NERVES

This specimen displays the brain and major nerve pathways of the central and peripheral nervous systems. The central nervous system is composed of the brain and spinal cord, which are situated in the cranial cavity and the vertebral canal, respectively. The brain connects with the spinal cord through the foramen magnum at the base of the skull.

Nerve impulses from the whole body are transmitted continuously upwards to the brain via the spinal cord and cranial nerves to produce sensation. These same nerve pathways allow the brain to send signals to the body commanding it to move and instructing the organs to perform their functions.

There are 12 pairs of cranial nerves that emanate from the brain and 31 pairs of spinal nerves that extend from the spinal cord into an extensive branch of nervous networks. These nerves, together with the visceral nerves that regulate the activities of the thoracic and abdominal organs, constitute the peripheral nervous system. This is why you have the ability to perform such delicate tasks as holding a single hair, feeling a single grain of sand underfoot, or threading a needle (providing you can see it!).

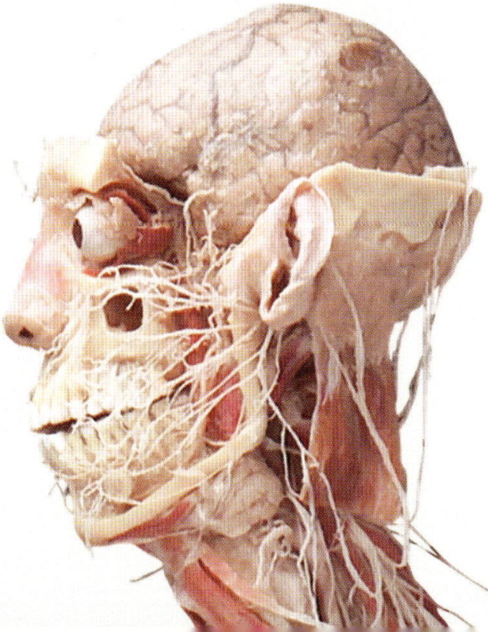

TRIGEMINAL NERVE
In addition to the 31 pairs of nerves that connect to the brain via the spinal cord, there are also 12 pairs of cranial nerves that emanate directly from the brain to supply commands and receive information from the structures of the neck and head. The trigeminal nerve is the fifth pair; it transmits sensations from the mouth, teeth, face, and nasal cavity and receives motor commands from the brain for chewing and assisting with speech. A part of this nerve is numbed when you go to the dentist.

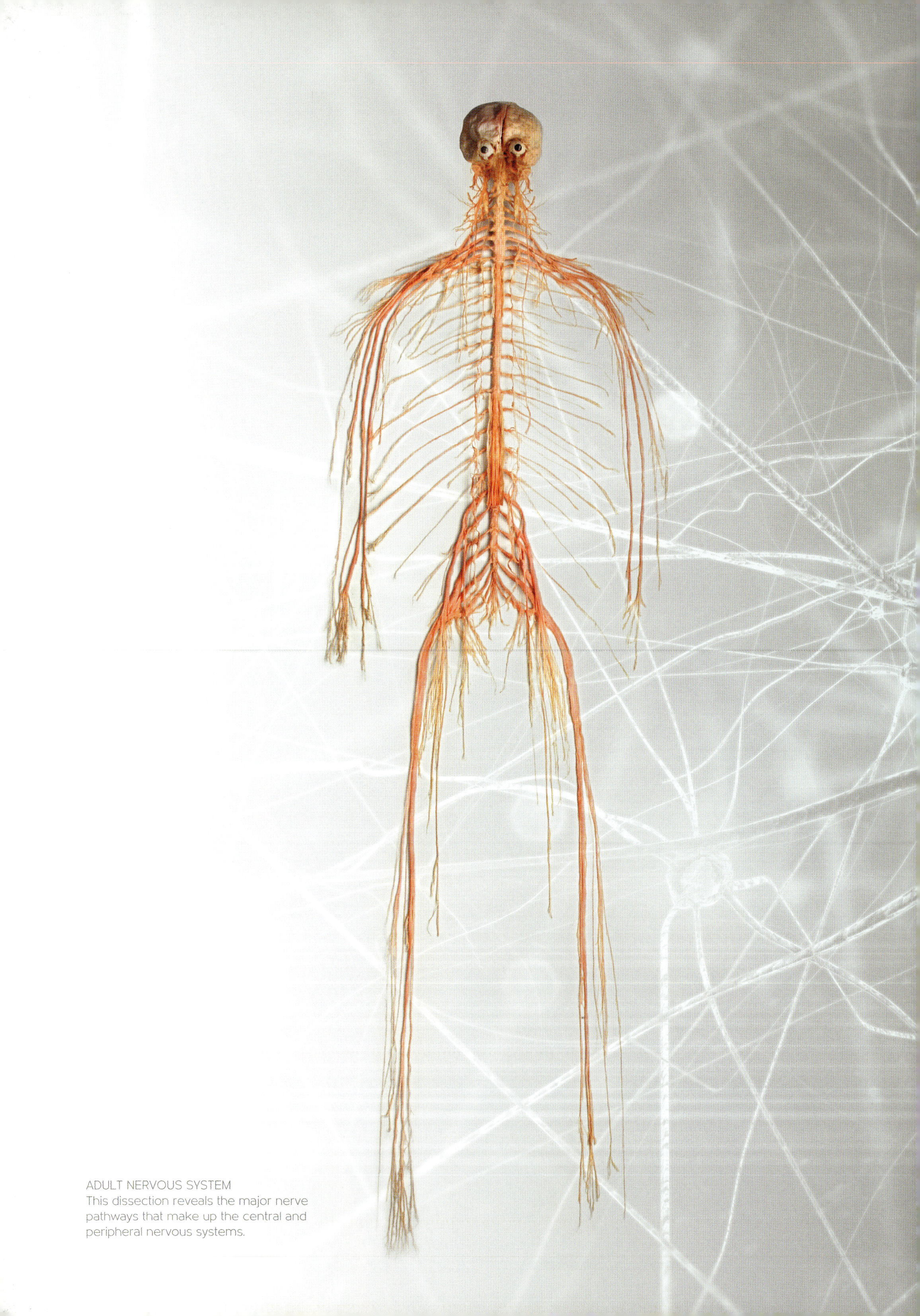

ADULT NERVOUS SYSTEM
This dissection reveals the major nerve pathways that make up the central and peripheral nervous systems.

THE MIND

Since ancient physicians began to study the human body, they have created a distinction between Brain and Mind, and this dichotomy still exists today. The Brain is a physical organ; it can be seen, dissected, and observed. The Mind, however–a collection of our memories, dreams, hopes, and anxieties–cannot be found in any examination or dissection.

Today, we scan the brain with sophisticated imagining devices, we monitor its electrical activity and trace the paths of its complex neural networks; we even divide up parts of the brain to point them to specific activities having to do with biological function, speech, and behavior. Yet, despite all our physical mapping of the Brain, we have yet to understand how all of its electro-chemical activity leads to consciousness–the seat of the Mind.

Our consciousness–a sense of self, of purpose, of meaning–is not located in one region of the Brain, yet it is fueled by the Brain's power. It is as though all those Brain regions and their respective parts conjoin to become something greater than a whole. They transcend the physical piece of the body called Brain to become that mysterious individual known as You.

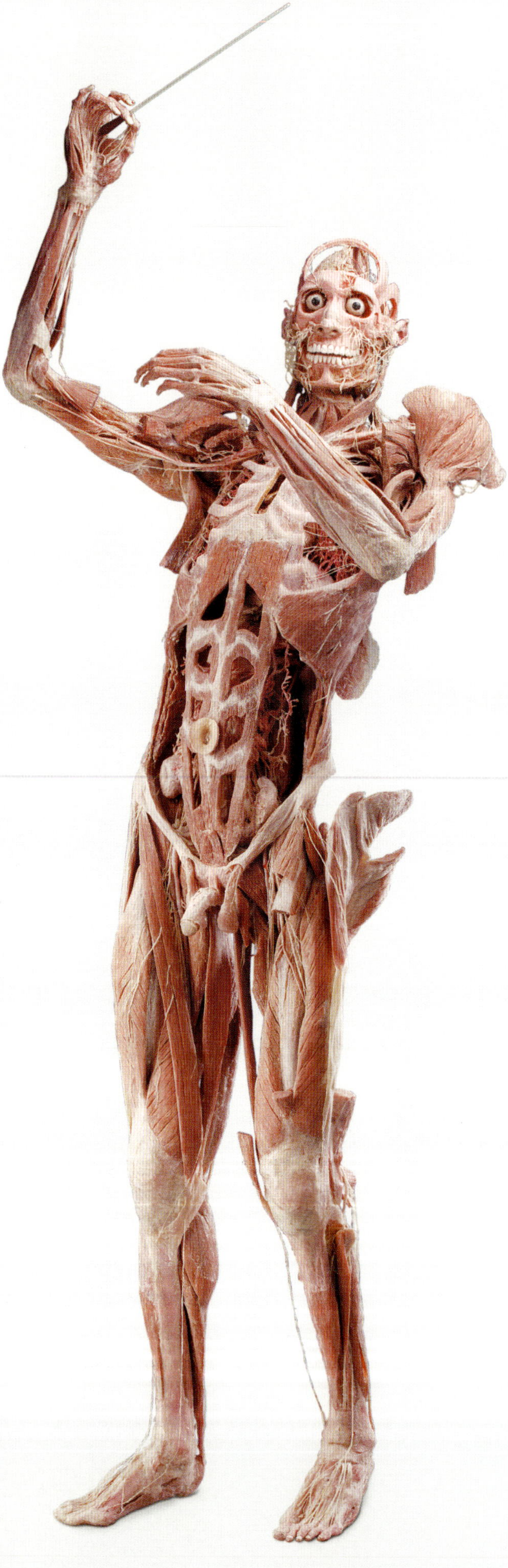

MOTOR CONTROL

This specimen displays the central and peripheral nervous system of the human body while also retaining the thoracic and abdominal walls and the sutures of skull. In this manner, we can see how the nerves inform the body wall and extremities as well as their inner pathways.

In this specimen, most of the cerebral and facial cranium has been removed to clearly display the sutures of skull, the meninges of brain, as well as the brain itself and the nerves deep in the face. The back of the specimen displays the sciatic nerve, the meninges of the spinal cord, the rootlets of spinal nerves, and the divergent sheaf of spinal roots at the base of the spinal cord that distribute to the lower extremities. This distribution of nerves is called the "cauda equina" or horse's tail because of its woven appearance.

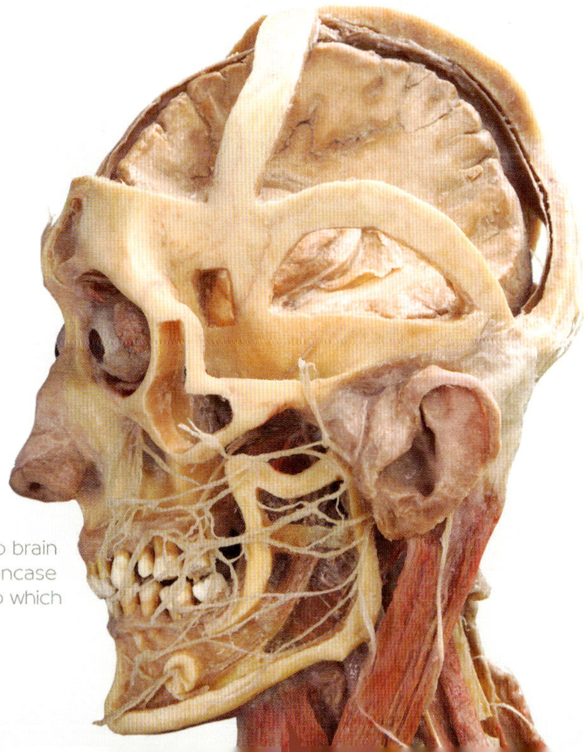

MENINGES AND FACIAL NERVES
This specimen reveals the deep brain structures and meninges that encase the brain as well as the extent to which facial nerves extend.

CEREBRAL VENTRICLES

There are four cavities in the brain, which are named ventricles. These ventricles are filled with cerebrospinal fluid (CSF) which plays a role in brain health and, like lymph, has nutritive and anti-infection functions.

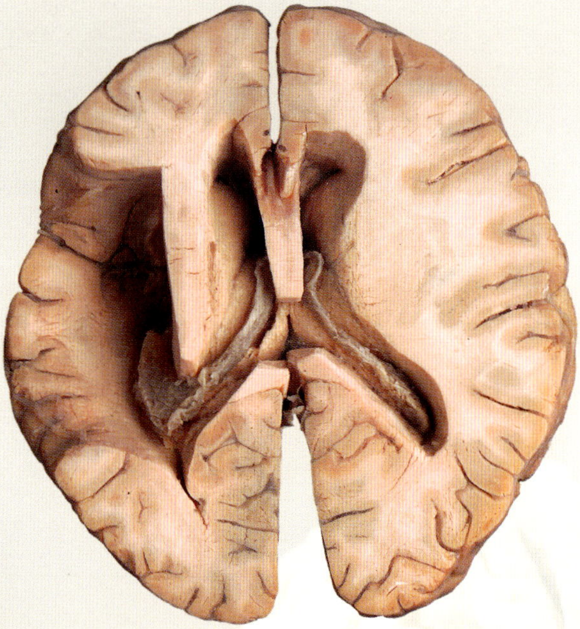

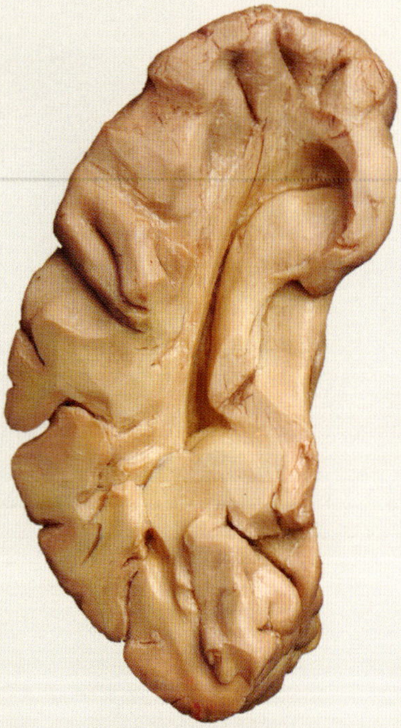

ENDOCRINE SYSTEM

The endocrine system produces hormones that deliver chemical messages to the cells, which affect everything from mood to behavior to heart rate to sexual function. Situated throughout the body, the organs of the endocrine system can be found in the brain, throat, abdomen, groin, and atop the kidneys.

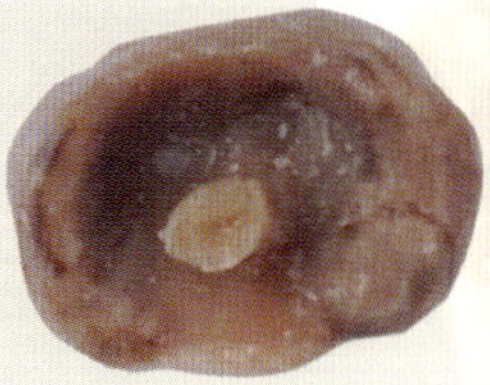

THYROID GLAND

A major organ of the endocrine system, the thyroid gland assists with regulating our metabolism, heart rate, and digestion. It also assists with brain growth in the developing fetus.

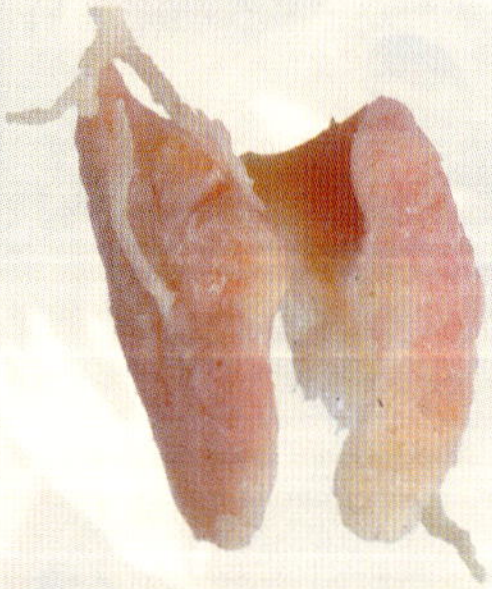

PITUITARY GLAND

Often referred to as the master gland of the endocrine system, the pituitary gland produces hormones that control many functions of other endocrine glands such as the adrenal and thyroid glands. Approximately the size of a pea, the pituitary consists of two lobes: the anterior (front), and the posterior pituitary (back).

INSULAR LOBE

Located deep within the brain, the insular lobe is believed to play a part in consciousness and homeostasis (keeping the body in a state of balance). Its functions include interpersonal relations, self-awareness, perception, and motor control. Some consider this the seat of the mind.

MENINGES
The brain and spinal cord are covered by three layers of membrane known as meninges. The outer layer, the cerebral dura matter, serves the principal function of delivering nutrients to the brain via arteries, veins and venous sinuses embedded within its thick, dense structure.

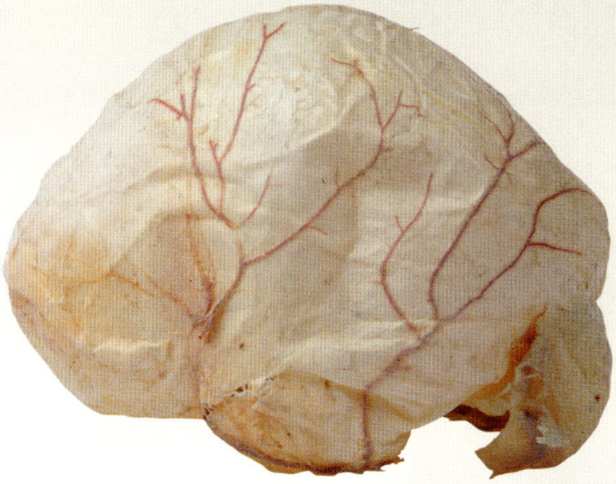

THE BRAIN
The brain is located in the cranial cavity, and consists of more than 100 billion nerve cells (neurons) and at least the same number of glial cells, which provide nutrition and support to the neurons' processes.

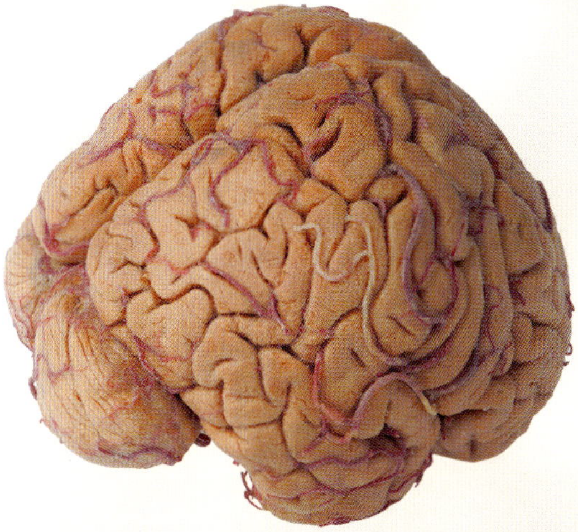

CEREBELLUM
The cerebellum lies under the cerebrum and behind the brain stem, and has extensive connections with the brain, the brain stem and spinal cord. The functions of the cerebellum include maintaining the equilibrium of the body, muscle tone, and smooth, coordinated muscle movements.

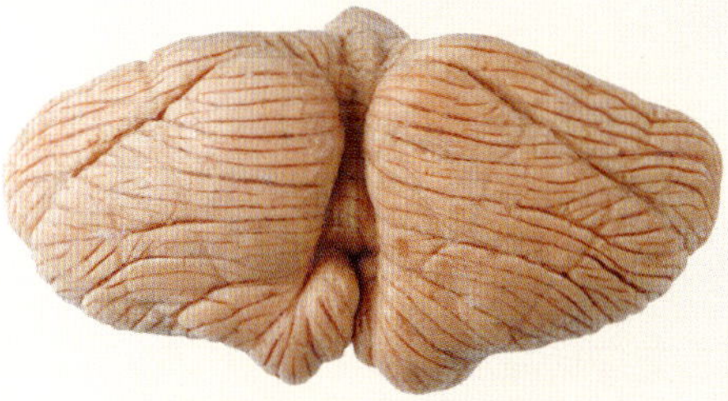

BRAIN STEM
The brain stem consists of the midbrain, the pons, and medulla oblongata. It contains the centers that control respiration, heart rate, and blood pressure. It also controls many reflex activities such as swallowing, coughing, and vomiting. The oldest part of the brain from an evolutionary perspective, it is sometimes referred to as the seat of the soul.

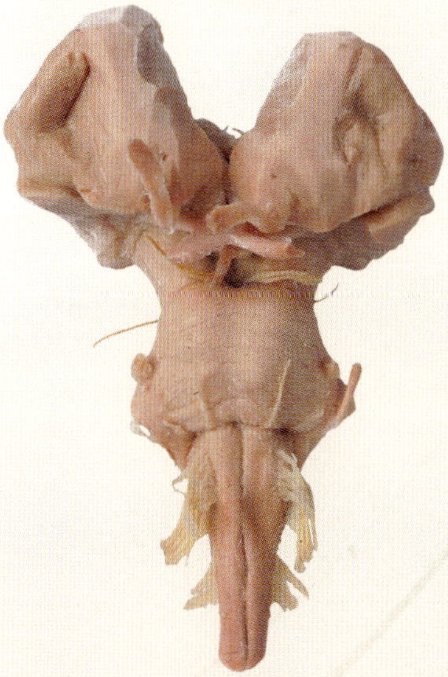

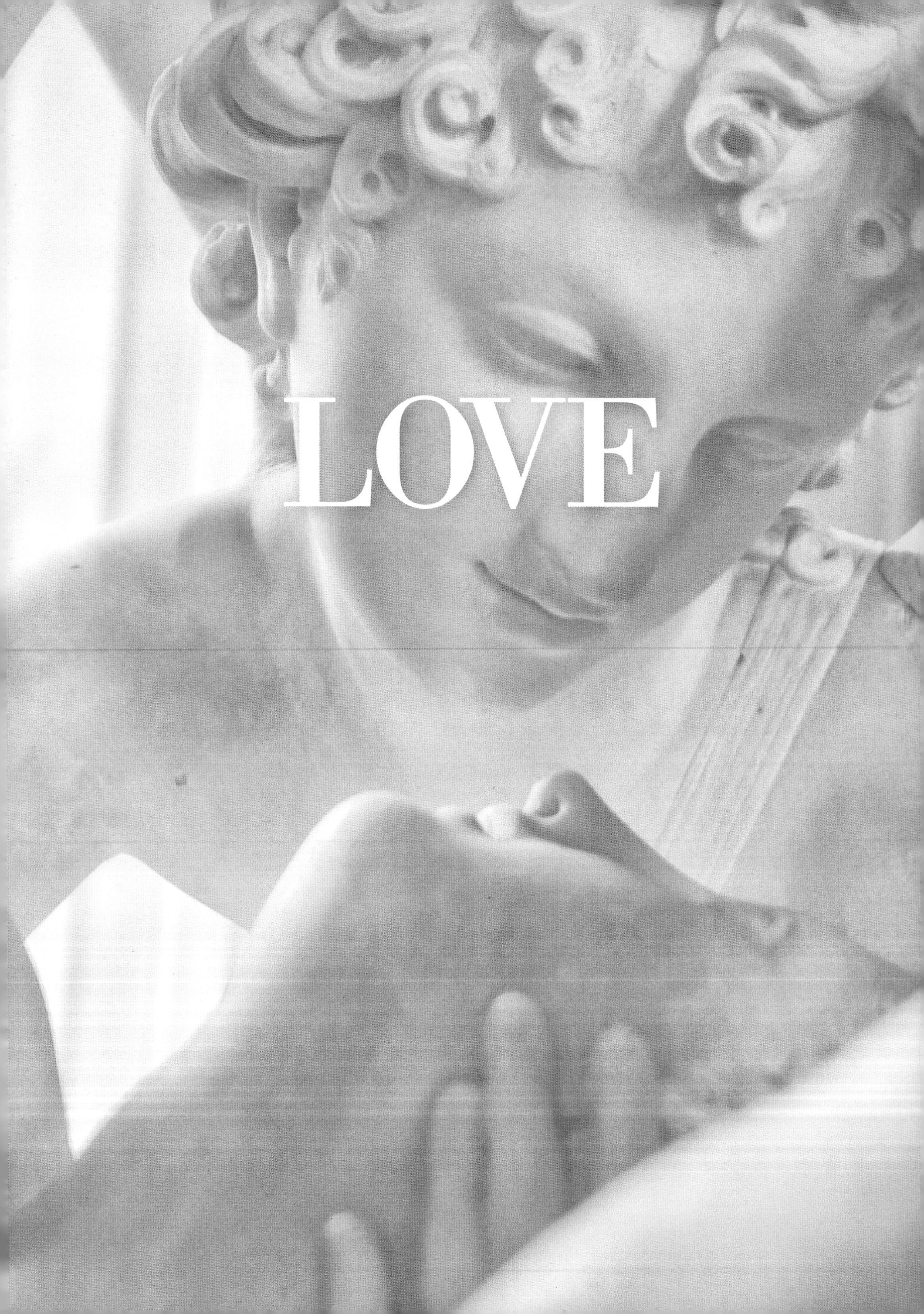
LOVE

ON LOVE

There is a spark of recognition. The eyes are familiar to you and you are drawn to them. It goes beyond attraction, beyond desire–these are ever present, essential to the species, but aren't important in this eternal moment. Rather, it's an opening, an invitation. "I have a place for you," it says, "for you above all others." And, it asks, "Will you join me?" Thus begins a journey of connection and separation, of celebrations and pitfalls, of communication in silent glances and smirks, of deep warmth even on the coldest nights. Alone, we are a heap of bone and blood, 11 types of atoms furiously struggling to stay assembled; but here, wrapped in the mystery of this connection, hunger and poverty fall away–for here, we are truly known.

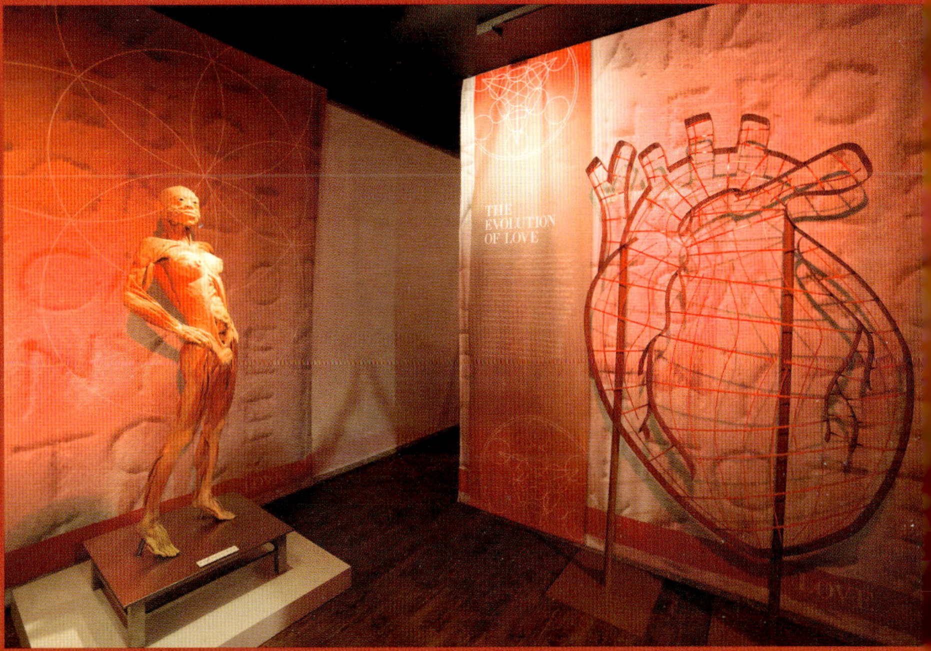

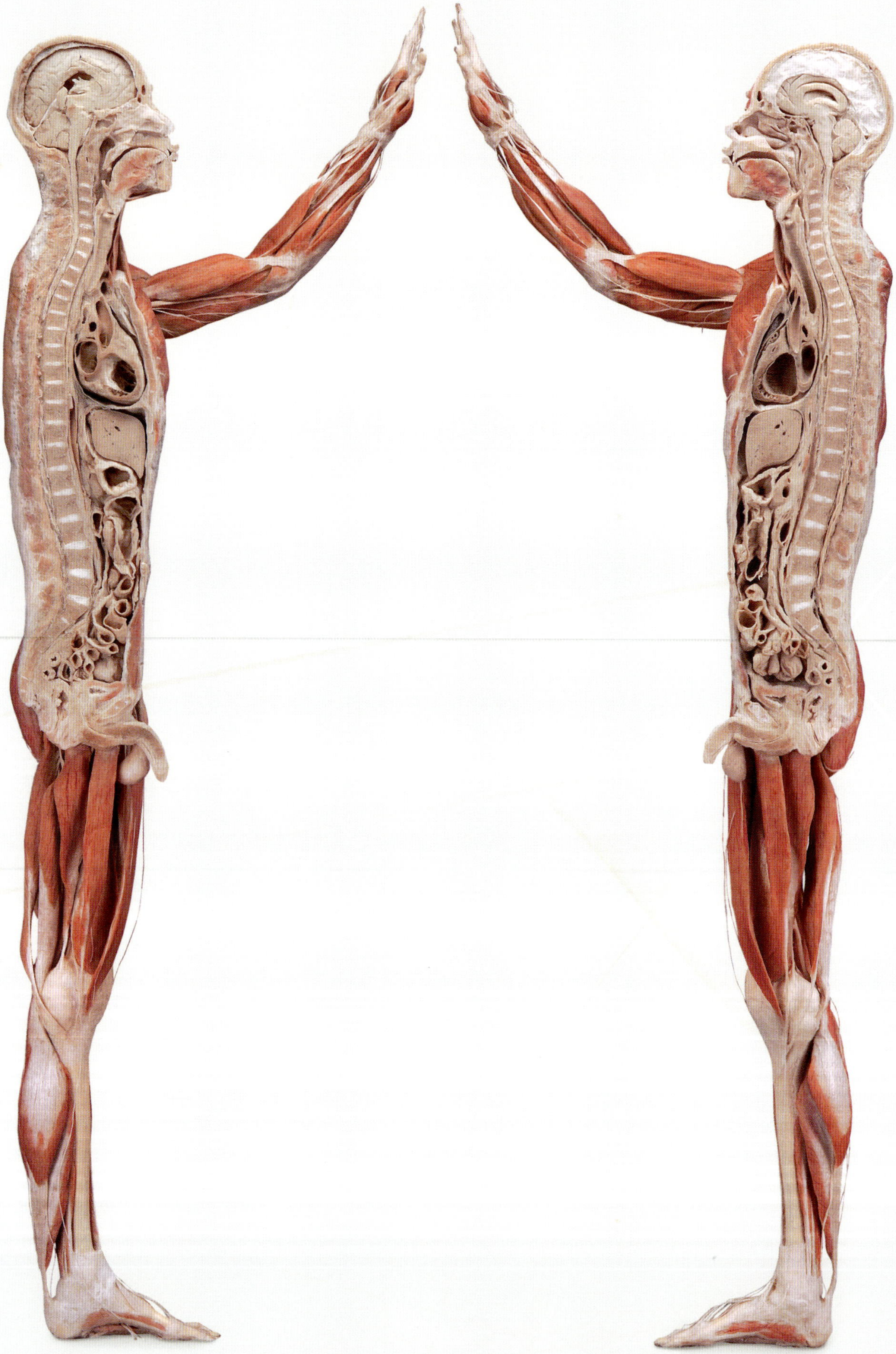

SYMMETRY + THE MALE BODY

This specimen helps us immediately understand the importance of symmetry to the external structure of the body. Symmetry of the body gives us balance in our constant battle against gravity and allows our motions to be fluid and efficient when walking, running or grasping objects. Symmetry of the face allows for depth perception and directional hearing. Within the body, however, there is less symmetry. For example: the left and right functions of the brain are different; the heart is not centered on the midline; and the lungs are different sizes.

The male reproductive organs, however, exhibit more symmetry, perhaps because of their importance in the preservation of the species. The male genital organs consist of the paired testes, epididymides, ductus deferentes, seminal vesicles, and the singular prostate gland and penis. The two testes, located in the scrotum, produce sperm and male sexual hormones which travel up one of two opposing epididymis glands to respective vas deferens ducts, which pass close to the opposing seminal vesicles that help produce seminal fluid. At the prostate gland the male symmetry unifies in order to move its seminal fluid into the ejaculatory duct and out of the body through the urethra in the penis.

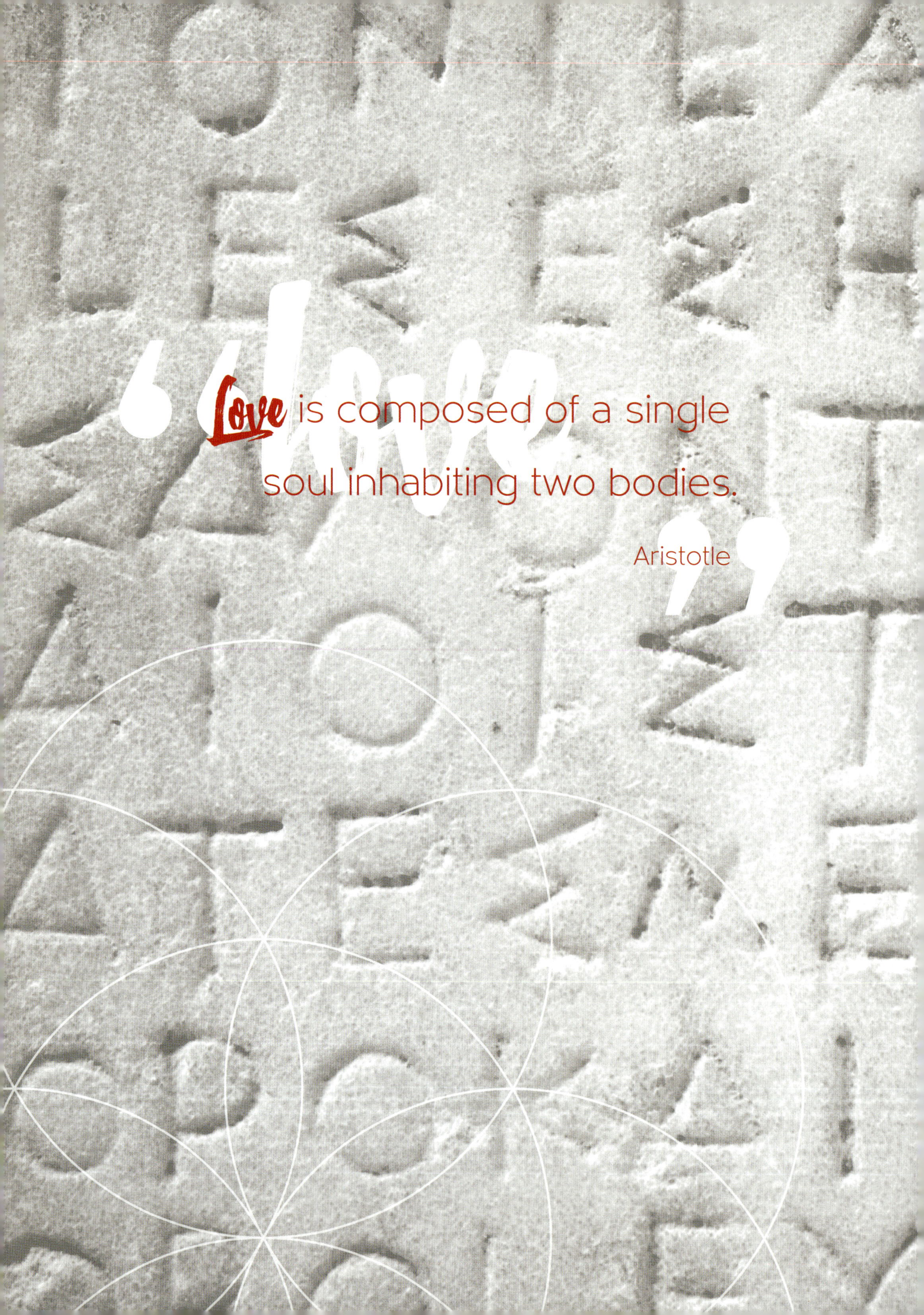

“Love is composed of a single soul inhabiting two bodies.”

Aristotle

THE EVOLUTION OF LOVE

Per the rules of Life, each species is required to reproduce in order to ensure its survival. For most organisms, this basic requirement is met in basic ways–reproduce and move on. For humans throughout history, however, an elaborate series of mating rituals have evolved to ensure we stay on the planet.

For humans, certain elements of biological attraction are believed to be hard wired from our time as primates. Symmetry of the face, a certain waist to hip ratio, and scent are powerful elements in those "love at first sight" immediate attractions; but they come to us from a time when we were first learning to walk on two feet. Physical attraction among the earliest humans indicated (unbeknownst to them) healthy genes that were most likely to produce offspring that would survive.

Love and commitment are not just constructs of today's greeting card, Hollywood machine either. Romantic love stemmed from the ingrained necessity to protect one's mate; partnering, which later turned into commitment, provided offspring with the best chance of survival.

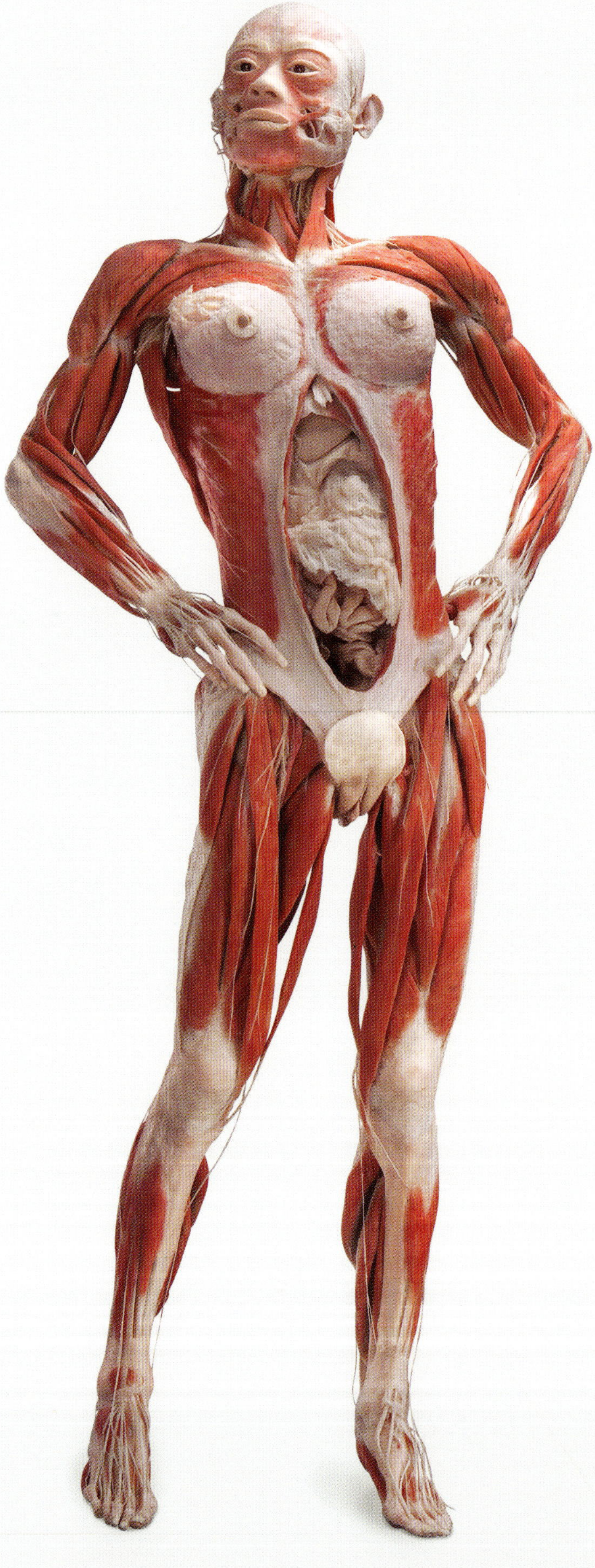

THE FEMALE BODY

This specimen mainly displays the superficial muscles and the female reproductive system. The abdominal wall of the specimen was cut from the medial line to show the digestive organs within the abdominal cavity (the liver, stomach and intestines) and the female reproductive organs (the uterus, ovaries and uterine tubes) within the pelvic cavity.

The uterus is the location for the development of the fetus. The uterine tubes connect with both sides of the uterus, which transmit the ovum (egg) and are the site of fertilization. The ovaries lie below the uterine tubes; they produce eggs and secrete the female hormones estrogen and progesterone. Because impregnation and fetal development occur within the female body, the principal female genital organs are protected in the pelvic cavity. Only the female pudendum and the accessory genital organs (mammary glands) lie on the surface of the body.

At fertilization, a single sperm enters into the ovum and the two combine into one cell–a "zygote"; this occurs in the fallopian tube. The ovum divides into two cells 30 hours after fertilization, then continues to divide once each 12 hours. The splitting zygote moves into the uterine cavity 72 hours after fertilization and embeds in the inner membrane of the uterus, where it continues to develop until the birth. This process from fertilization until birth lasts about 266 days.

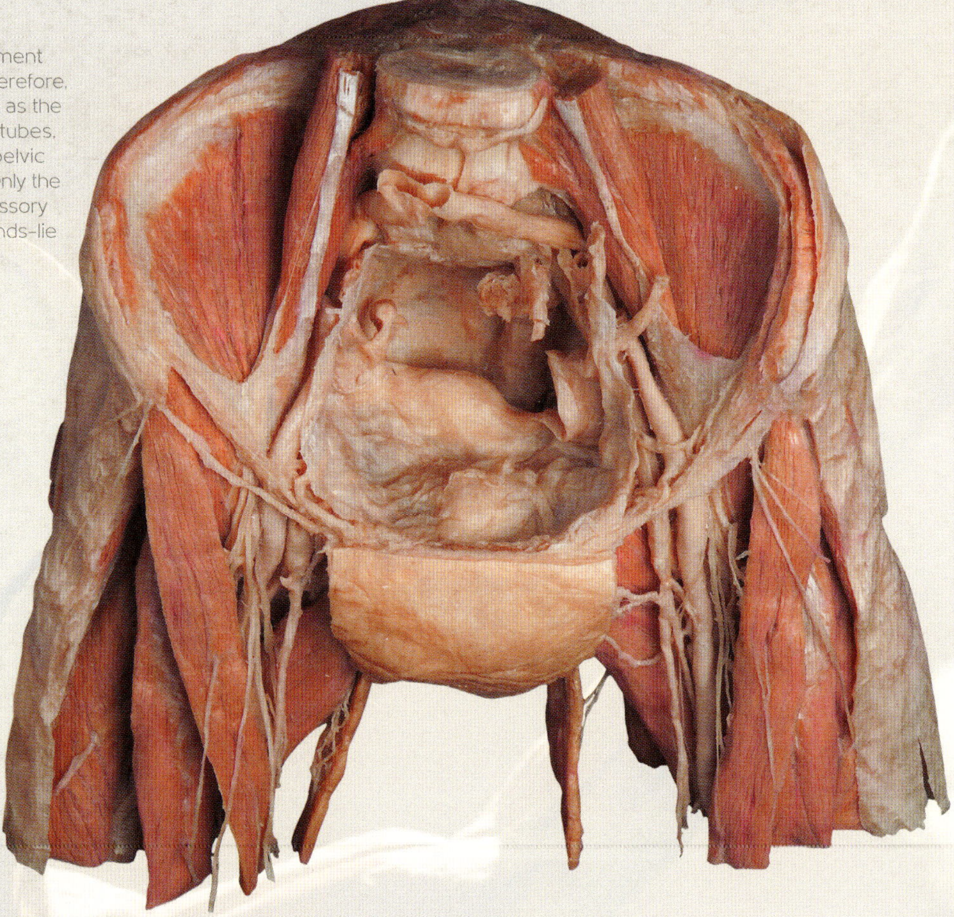

SUPERIOR VIEW OF
FEMALE PELVIC CAVITY
Impregnation and fetal development occur within the female body; therefore, the principal genital organs such as the uterus, the ovaries, the uterine tubes, and vagina are situated in the pelvic cavity for maximum protection. Only the female pudendum and the accessory genital organ–the mammary glands–lie on the surface of the body.

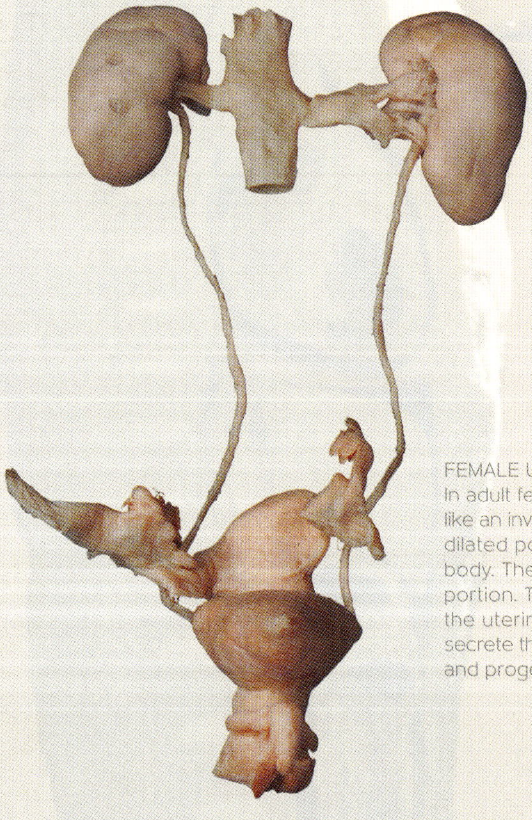

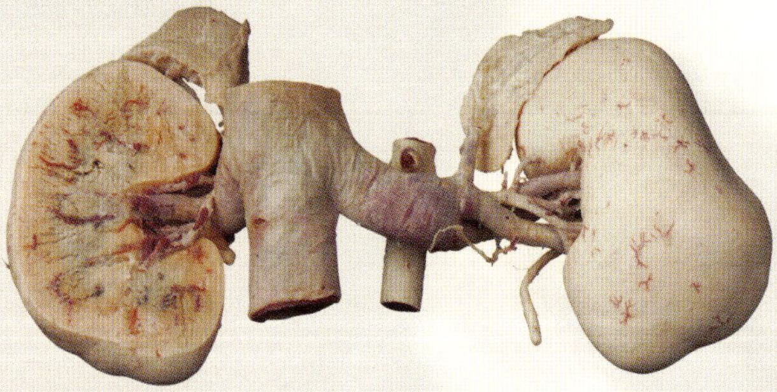

FEMALE UROGENITAL SYSTEM
In adult females, the uterus is shaped like an inverted pear, of which the upper dilated portion is termed the uterine body. The fetus mainly grows in this portion. The ovaries, which lie below the uterine tubes, produce eggs and secrete the female hormones estrogen and progesterone.

KIDNEYS
Kidneys are the principal organs of the urinary system. The kidneys remove certain metabolic wastes and excess water from blood in the form of urine.

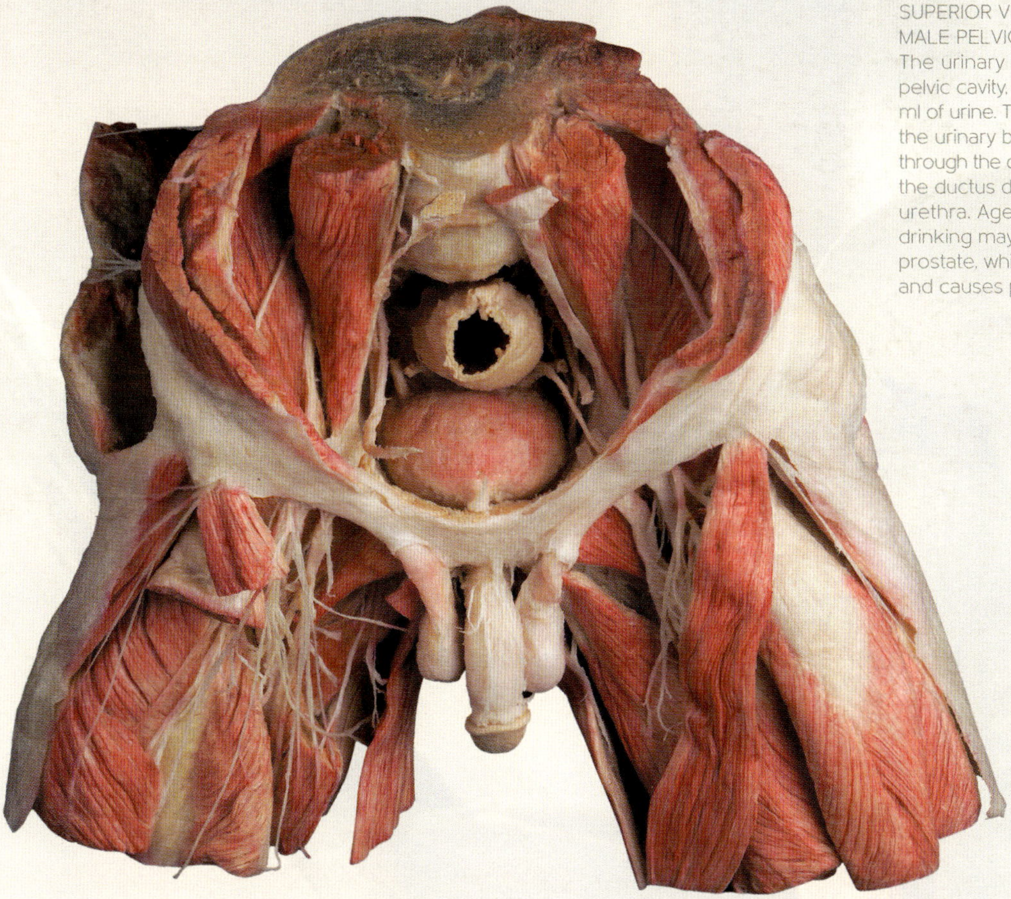

SUPERIOR VIEW O
MALE PELVIC CAVITY
The urinary bladder is situated in the pelvic cavity, which can store about 500 ml of urine. The prostate gland lies below the urinary bladder. The urethra passes through the center of the prostate, where the ductus deferens connects with the urethra. Age, smoking, and excessive drinking may cause enlargement of the prostate, which oppresses the urethra and causes problems in urination.

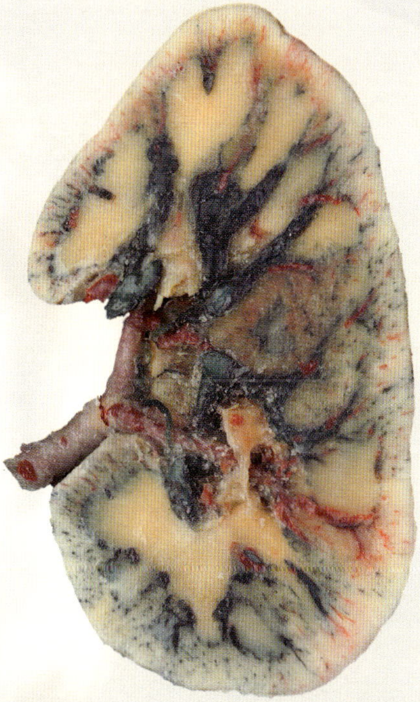

KIDNEY WITH CORONAL CUT
The kidney is divided into two portions: the cortex and the medulla. The renal cortex lies near the periphery of the kidney, while the medulla is deep in the cortex. The cortex is rich in blood vessels, which can filter blood and produce urine. The medulla reabsorbs useful substances from the urine and then expels the concentrated urine into the urinary bladder.

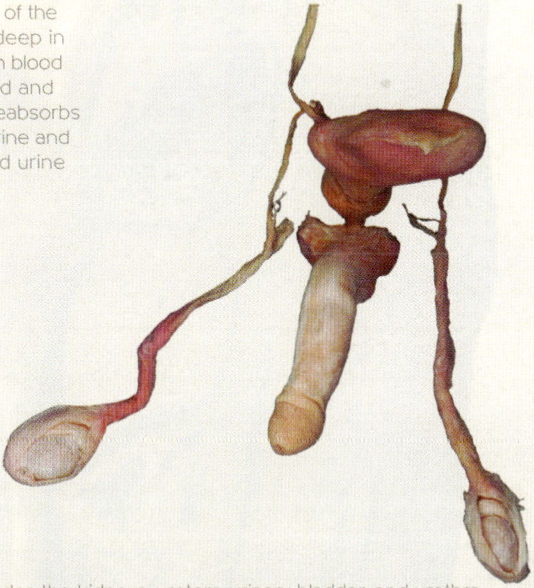

MALE UROGENITAL SYSTEM
The male urinary system includes the kidneys, ureters, urinary bladder, and urethra. The male reproductive system consists of internal genital organs (the testes, the epididymis, the ductus deferens, the seminal vesicles, and prostate) and external organs (the scrotum and penis). The testes are housed in the scrotum and produce sperm and male sexual hormones. Because sperm production occurs at a lower temperature than body temperature, the testes are most effective outside of the main body cavity.

BEGINNINGS

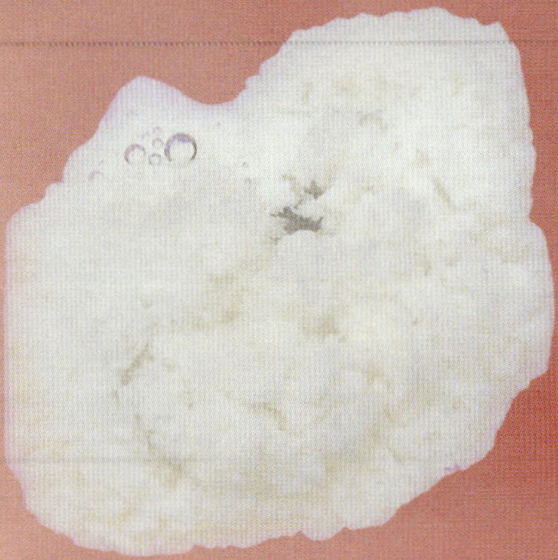

CHORIONIC SAC
Shortly after fertilization, the developing embryo forms the chorion–a membrane that encloses it and allows for transfer of nutrients from the mother.

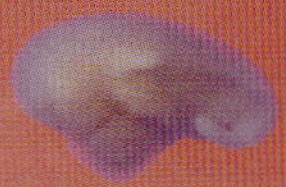

EMBRYO - 4 WEEKS
The cluster of cells (blastocyst) attaches to the lining of the uterus.

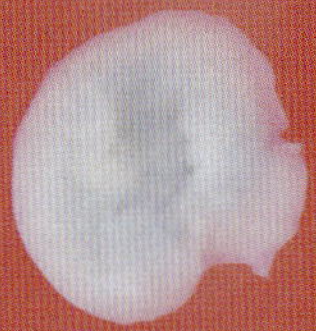

EMBRYO - 5 WEEKS
The placenta begins to form. The brain and spinal cord begin to form.

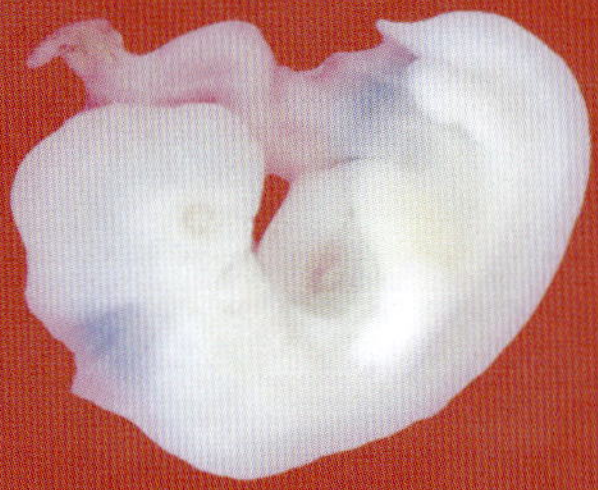

EMBRYO - 6 WEEKS
The tissues that will form the heart begin to beat. The heartbeat can be detected during an ultrasound exam.

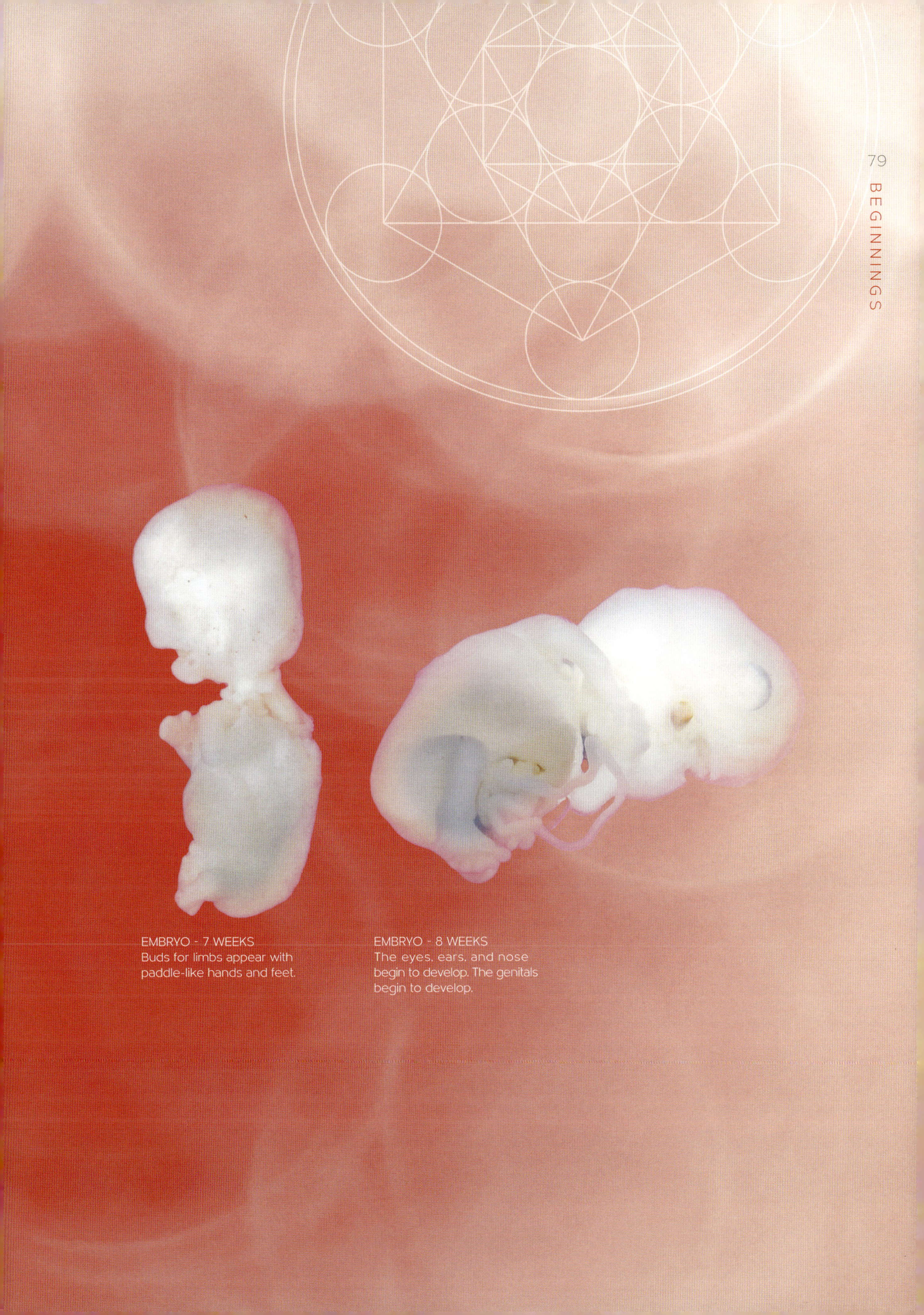

EMBRYO - 7 WEEKS
Buds for limbs appear with paddle-like hands and feet.

EMBRYO - 8 WEEKS
The eyes, ears, and nose begin to develop. The genitals begin to develop.

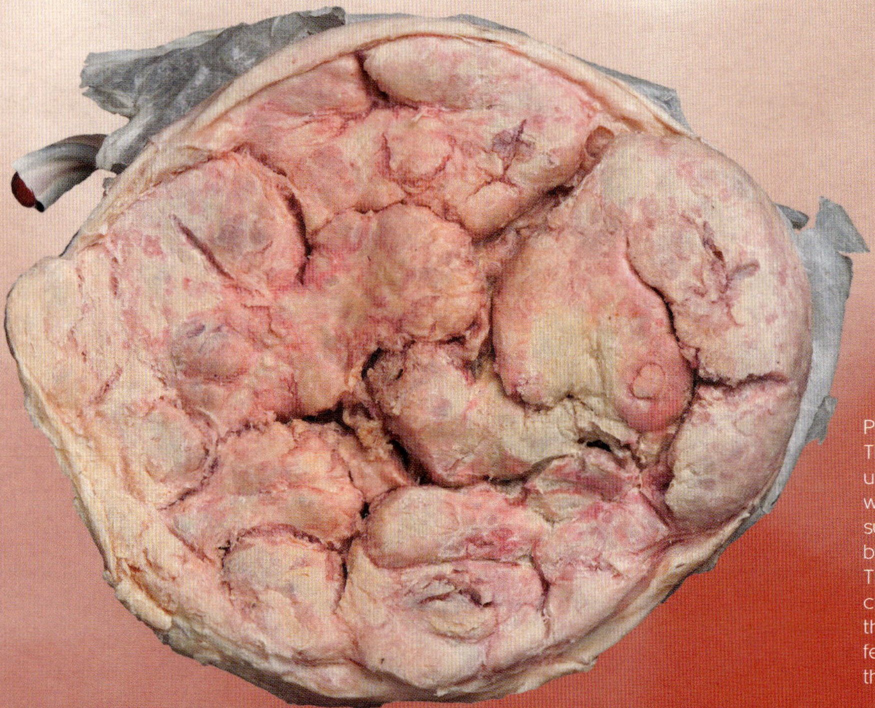

PLACENTA
The placenta is divided into the fetal and uterine sides. The uterine side connects with the uterine wall, which has an uneven surface. The fetal side of placenta is covered by a smooth and transparent membrane. The veins and arteries of the umbilical cord distribute in a radial pattern from the center to peripheral regions. The fetus absorbs nutrition and oxygen from the mother via the placenta.

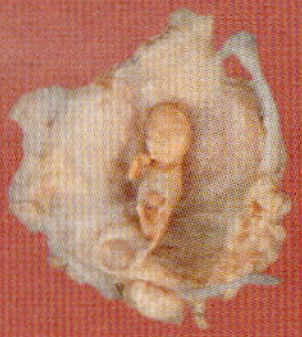

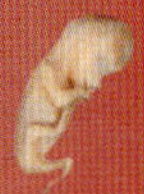

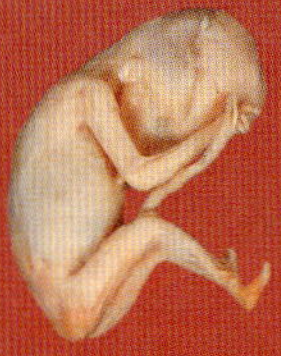

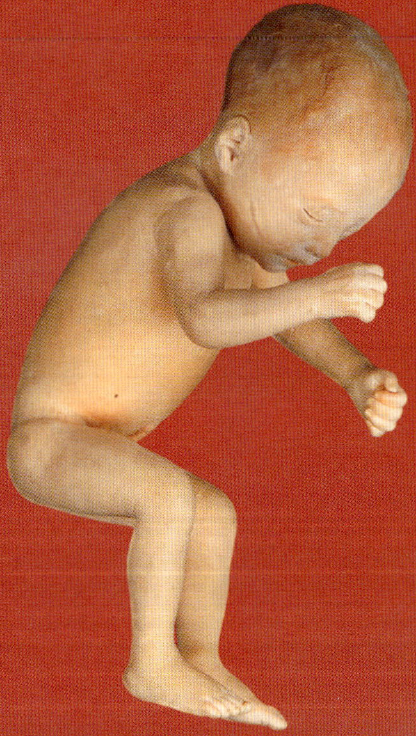

FETUS - 8 WEEKS
At eight weeks, the embryonic stage is complete and fetal development begins.

FETUS - 9 WEEKS
Weight gain begins, eyelids are formed.

FETUS - 13 WEEKS
As the first trimester of development concludes, fingerprints have developed.

FETUS - 32 WEEKS
The lungs are maturing and the skeleton has completely formed.

IN UTERO

In mother's womb
Embryo grows
Into fetus before pushing
The world's door open
To play the Game again
Continuing traditions
Or creating new ones

The cycle of Life
Ever unbroken
Is redrawn
With every new breath.

-Anonymous

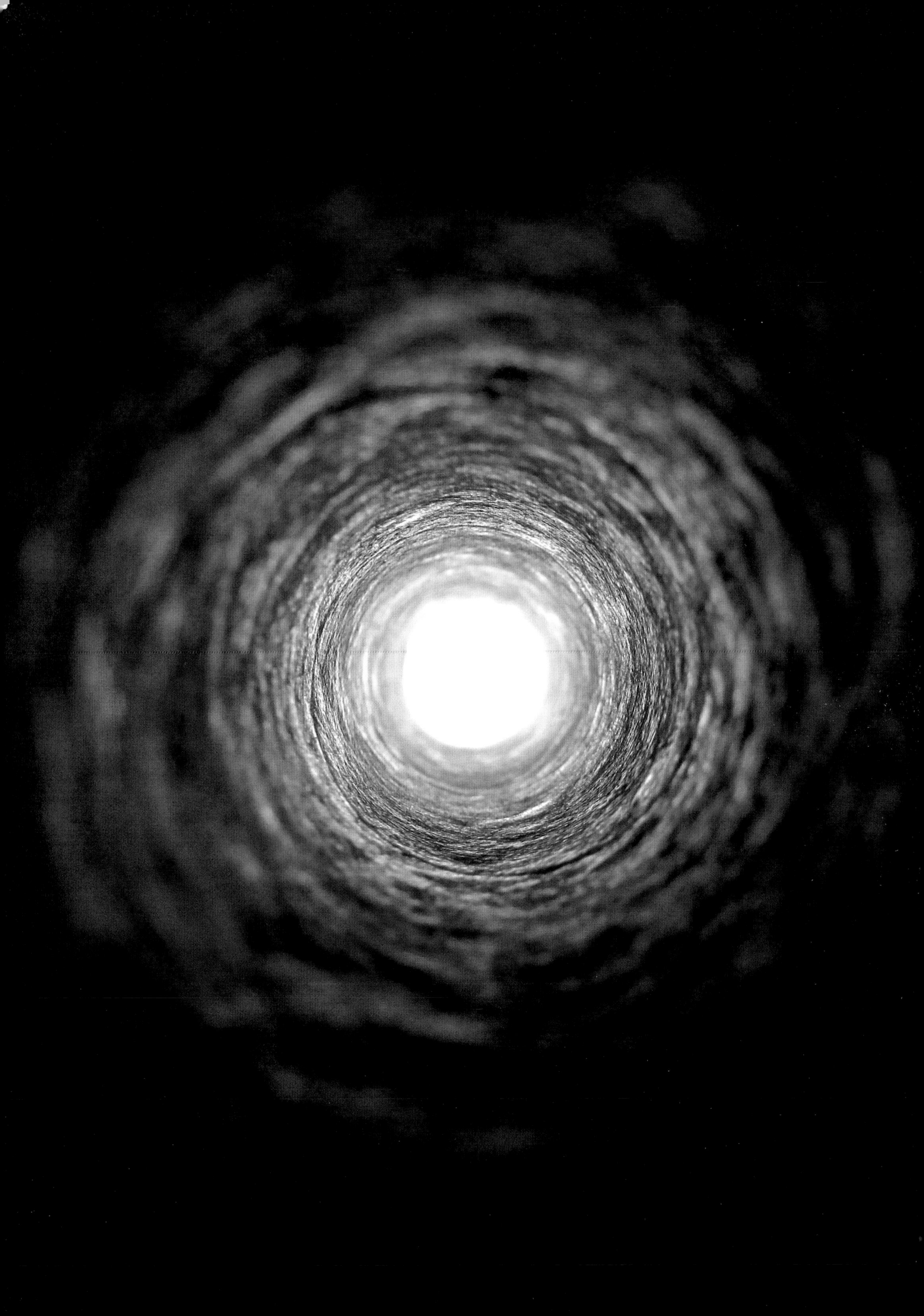

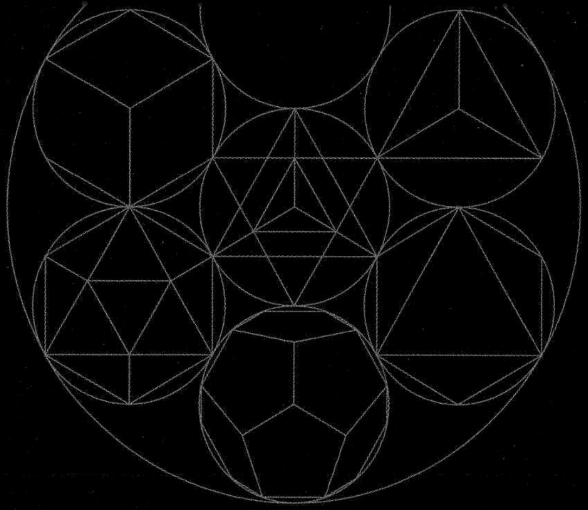

THE LAST FRONTIER

"The undiscovered country, from whose bourn no traveller returns"

Shakespeare, *Hamlet*

Where do we go after death? What do we leave behind?

We avoid the thought at all costs, but there will come a time (hopefully far into our future) when our bodies–these miraculous compositions of cells, organs, bone, flesh and blood–will be no more.

We call it the last frontier, the journey from this life to...where exactly? Into the ground, to ashes, to heaven, or maybe you have somewhere else in mind. According to one poll, 51% of the world believes there is some kind of life after death. Do you?

As you follow this brief transition hallway, take a moment to reflect on what lies beyond 'this mortal coil.' Once we shed our skin do we simply return to the ground, or do we take up a place among the stars? And what of our loved ones who remain?

Death is a sobering reality that brings the intoxication of life into full focus.

Best of luck on your journey, we'll see you on the other side...

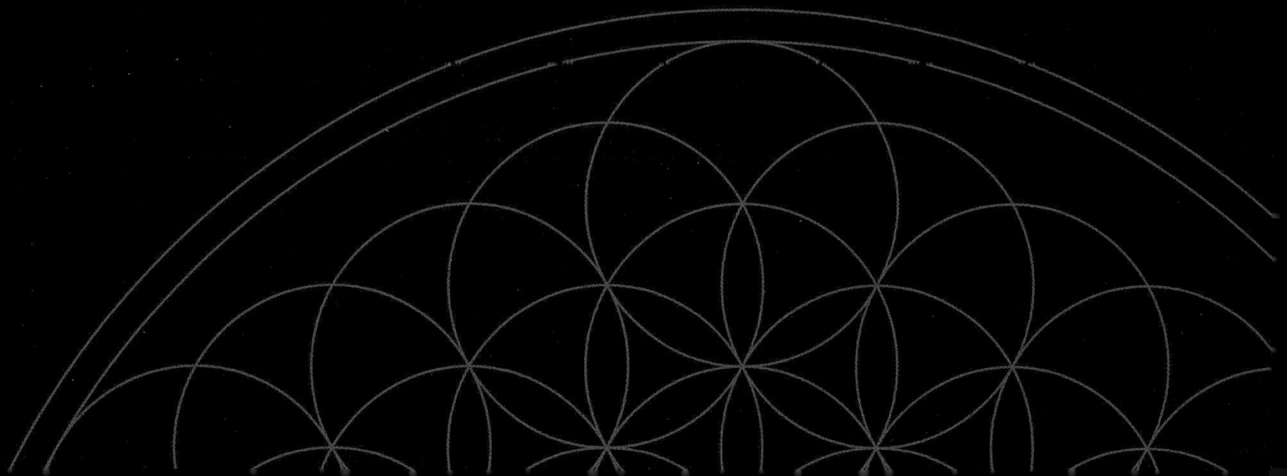

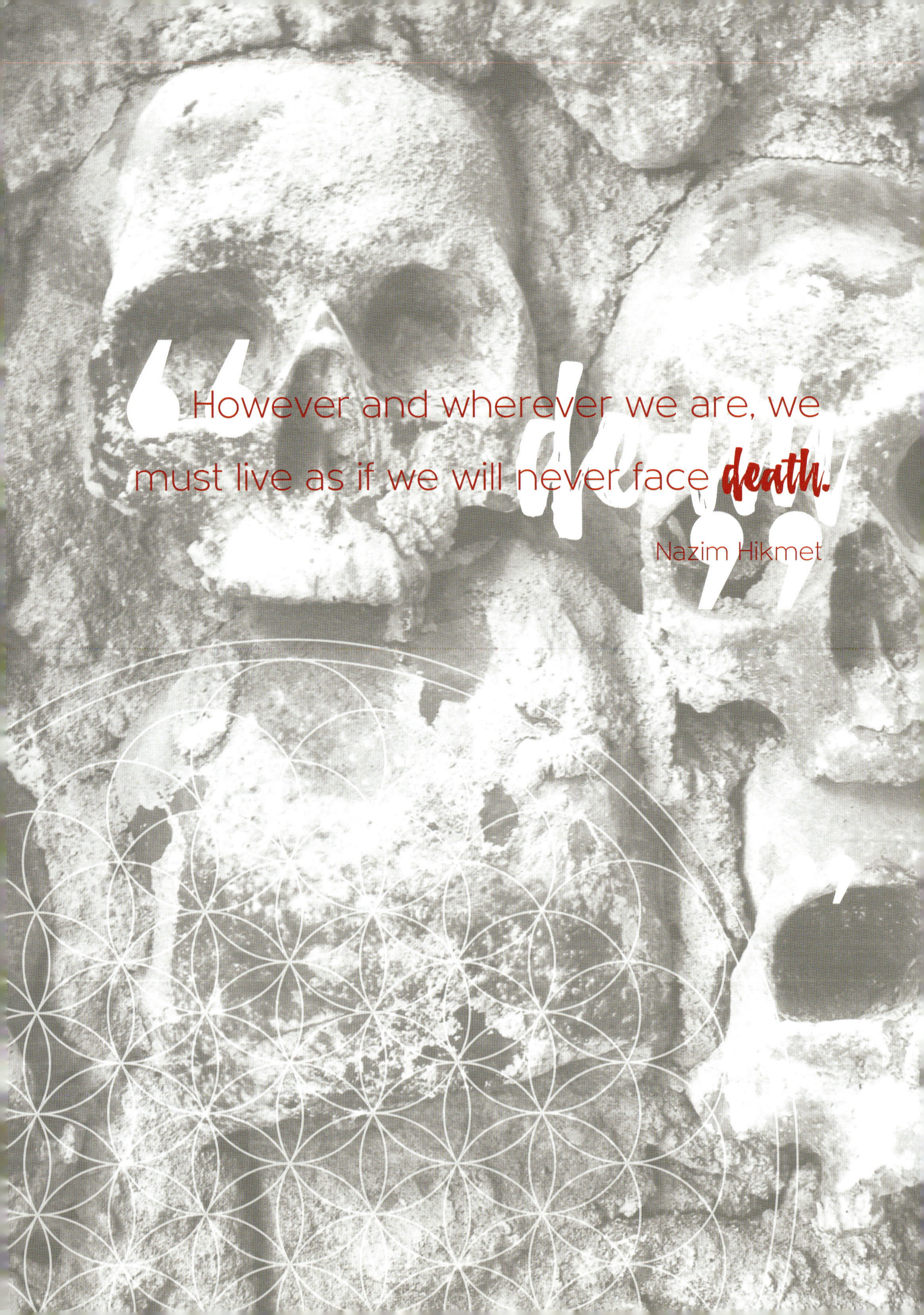
However and wherever we are, we must live as if we will never face death.
Nazim Hikmet

WHAT BECOMES OF US?

When we go into the ground–or turn to ash–at the end of our time, the organic material that has made up our bodies breaks down into its component parts, into the very atoms that we have held together for so long. For such a complex, and ultimately unexplainable system, it is surprising that only 11 types of atoms account for more than 99 percent of who we are. These are: Oxygen (65%), Carbon (18.5%), Hydrogen (9.5%), Nitrogen (3.2%), Calcium (1.5%), Phosphorous (1%), Potassium (0.4%), Sulfur (0.3%), Sodium (0.2%), Chlorine (0.2%), and Magnesium (0.1%).

The physical reality of death is quite straightforward; but it is also a mystery for those left behind. Cultures have developed elaborate rituals to honor those who have died, and while many mourn the loss of loved ones, other cultures find it a time for celebration–the New Orleans Jazz funeral is an example of this, as it is in eastern Indonesia where funeral celebrations can last for weeks. Other cultures keep the memories of their dead alive by turning their remains into colorful beads (South Korea), dressing them and placing a lit cigarette in their lips (Philippines), and burying them in "fantasy coffins" that most represent who they were while alive (Ghana)–an airplane pilot might be buried in a large plane coffin, a fisherman in a giant tuna.

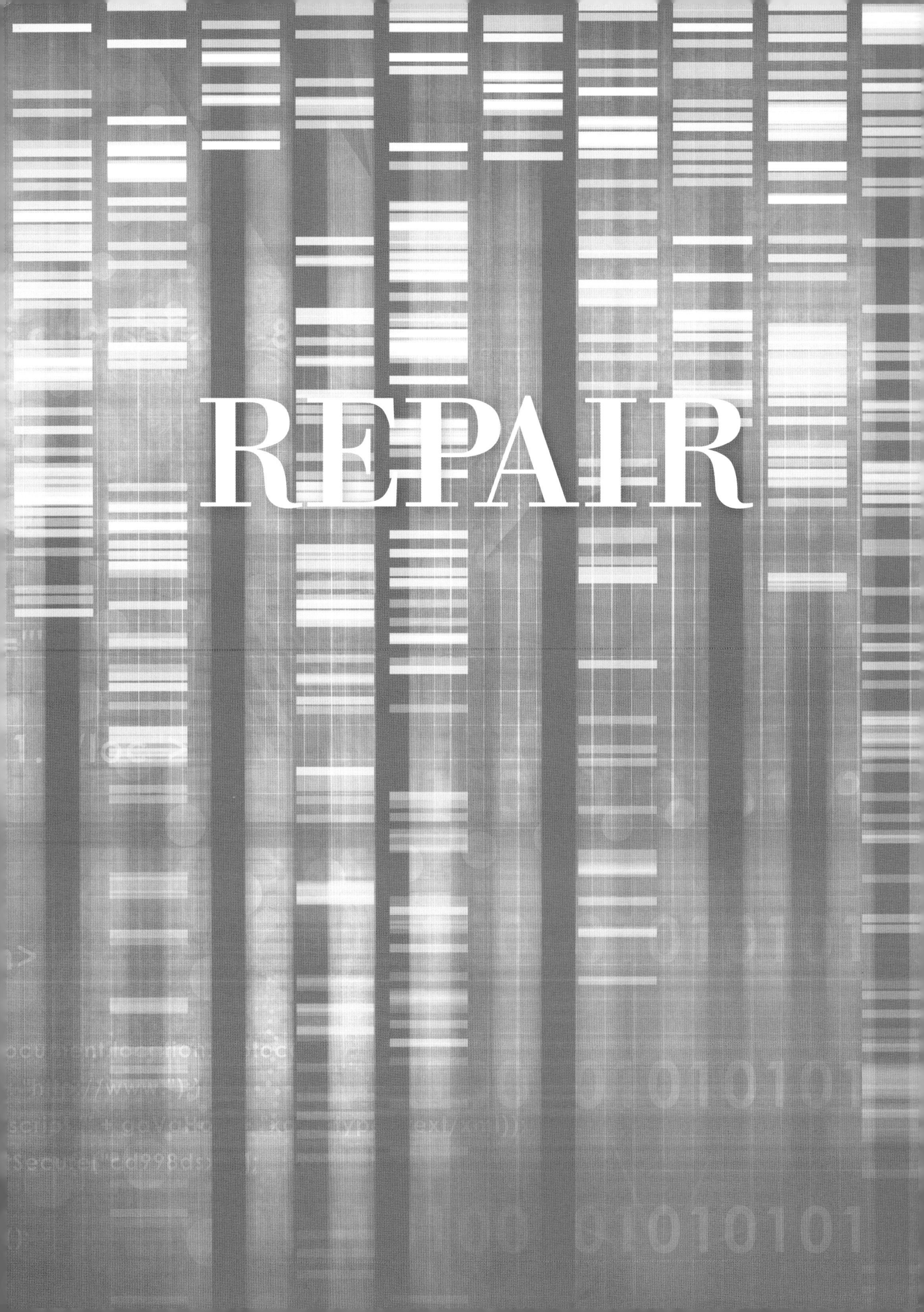

REPAIR

MEDICINE: YESTERDAY, TODAY, AND TOMORROW

This is a remarkable time for Western medicine. Worn-out joints are easily replaced, more than 100,000 people receive new aortic heart valves each year, and surgeons operate on tiny regions of the brain without affecting the rest of the organ. These are just a few examples of the sensational progress that medicine and surgical procedures are making.

In general, we are living longer and enjoying our health–and our bodies–much later into life.

At the same time, we are growing more aware of how lifestyle choices and the foods we eat can have a positive or adverse impact on our health. Yoga, meditation, and acupuncture are on the rise. These non-Western approaches to medicine–centered around a mindfulness of the body and care of the soul–have been around for millennia. Today, as these instinctual, harmonious methods infuse the more direct, sometimes invasive, practices of the West they are honoring the past as they help to create the medicine of the future.

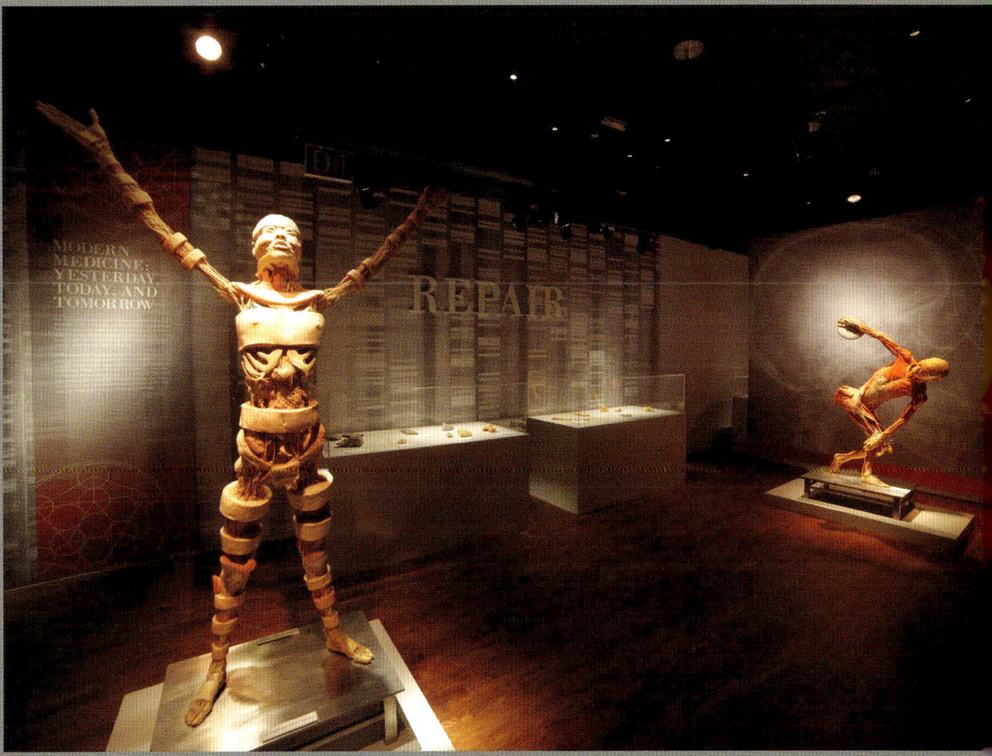

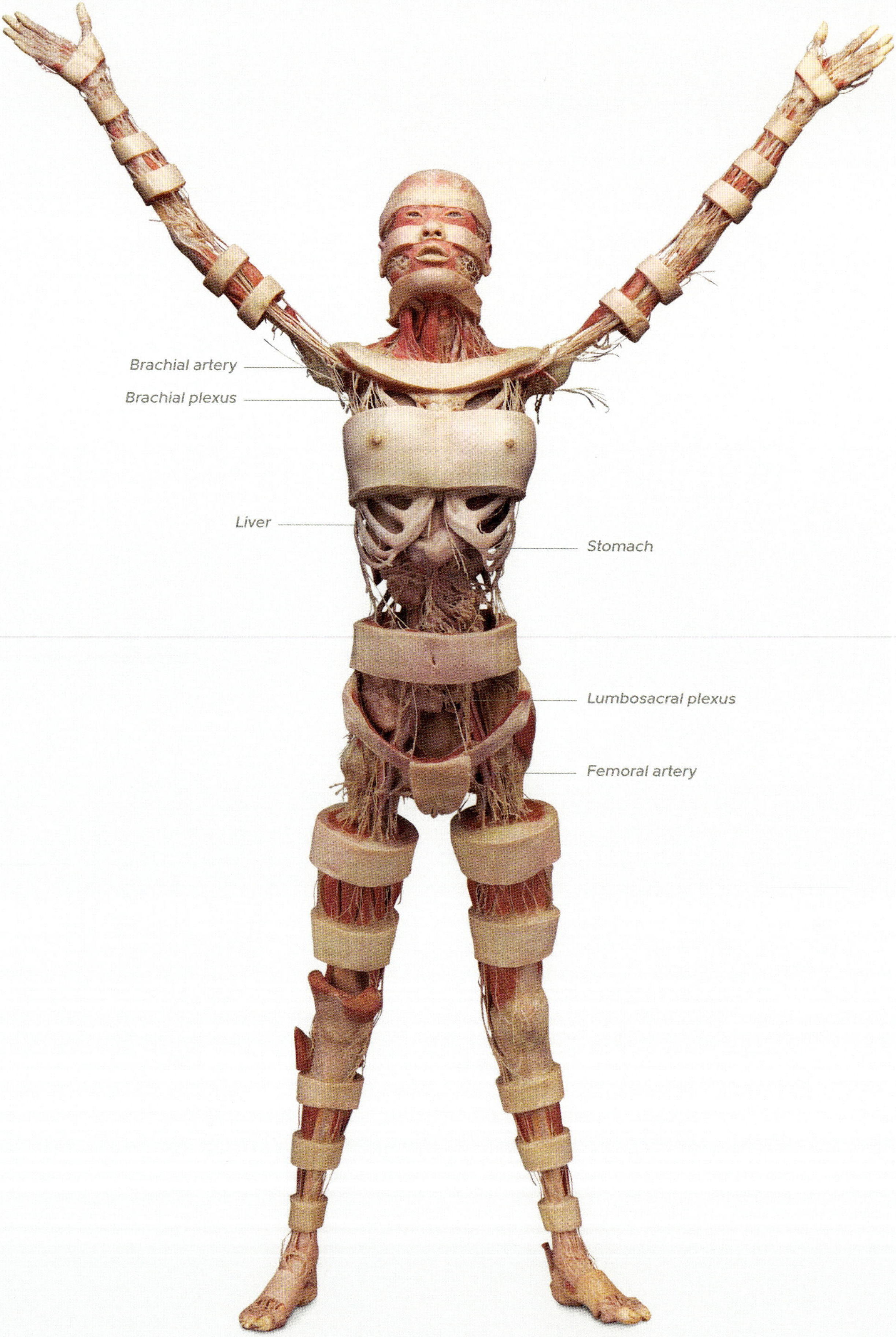

Brachial artery
Brachial plexus
Liver
Stomach
Lumbosacral plexus
Femoral artery

VITRUVIAN WOMAN

Inspired by Leonardo da Vinci's Vitruvian Man drawing, this specimen provides an insightful view into the layered nature of the body's composition and offers an excellent review of many of the body systems on display in this exhibition. In several areas, including the chest, abdomen, legs, and arms, you can see the how adipose tissue lies justbeneath the three layers of skin and above the skeletal muscle. At the extremities the tendons that move the fingers and toes are also clearly visible.

This specimen also provides an excellent view of the pathways that nerves and blood vessels take through the body as well as the areas in which they come together for efficient passage through the joints, such as under the arm and at the wrists. When looking under the arm, the brachial artery, which supplies blood to the arm is clearly visible as is the brachial plexus where the nerves that distribute to the arm gather. In the same manner the femoral artery and lumbosacral plexus that supply blood and nerves to the lower limbs can be seen near the pelvis and hip area.

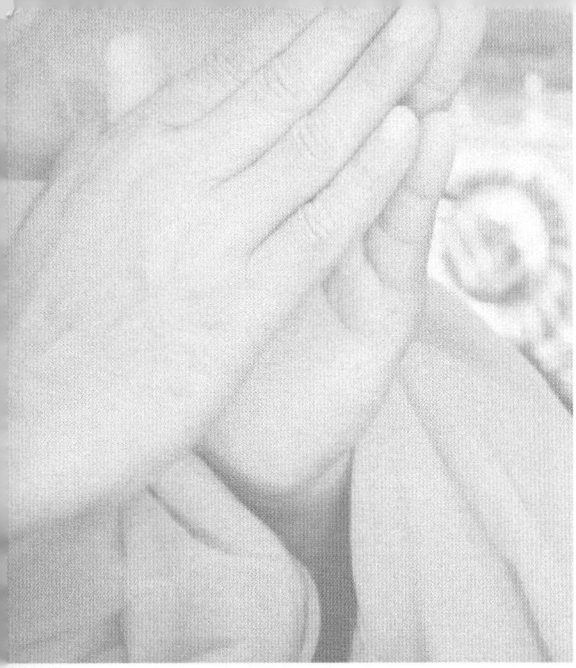
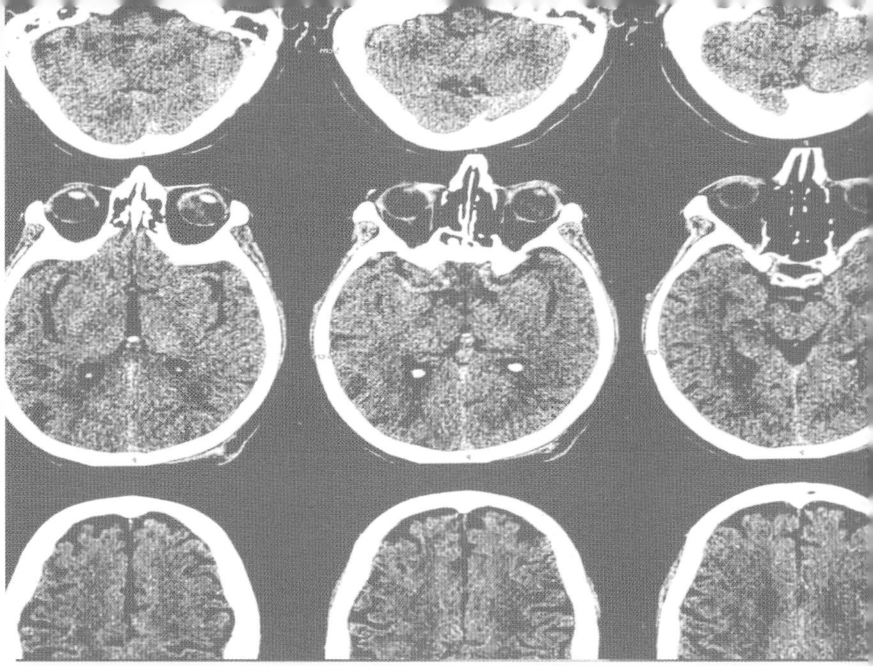
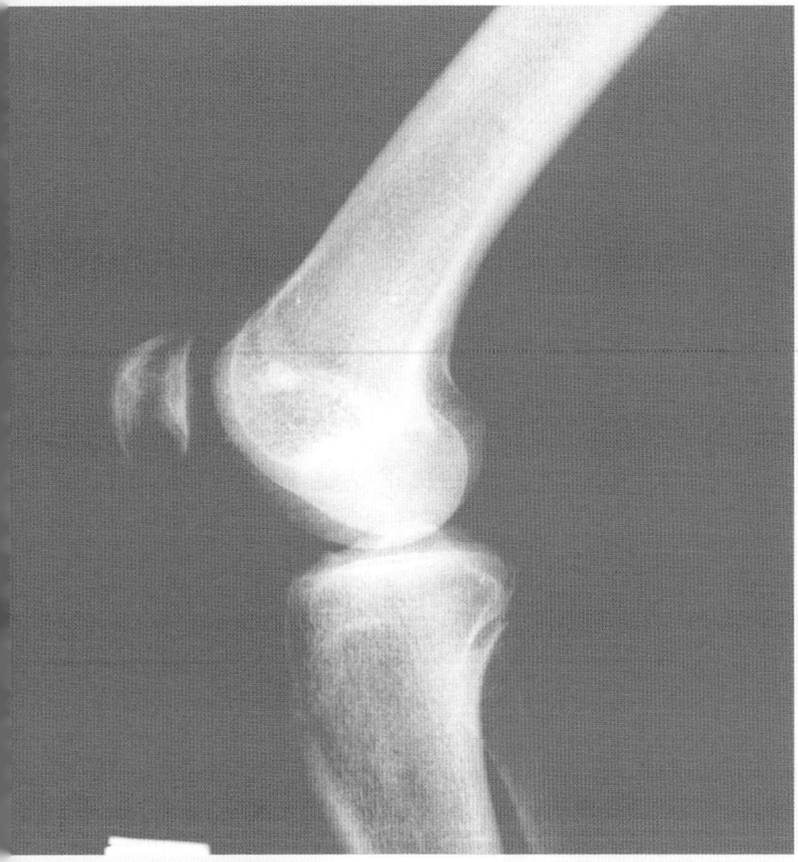
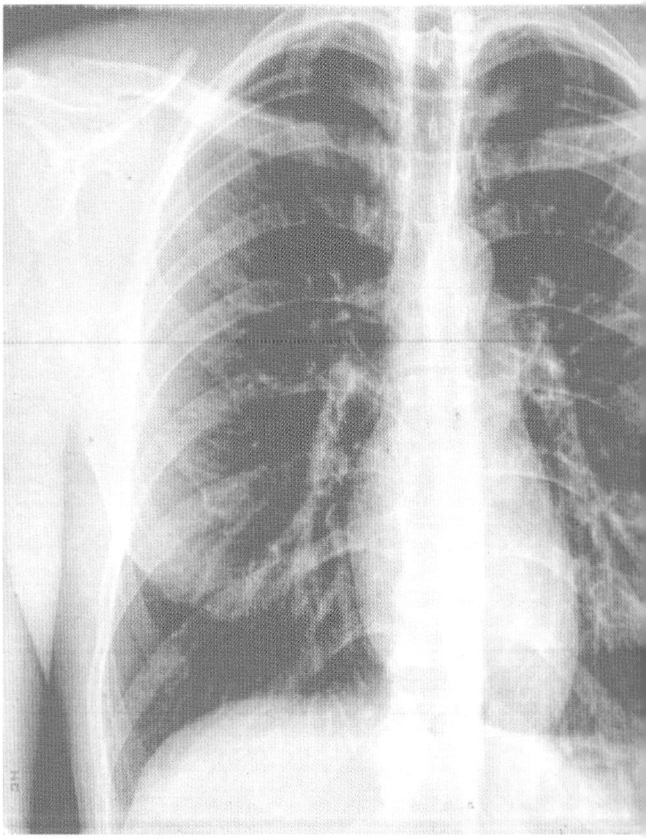
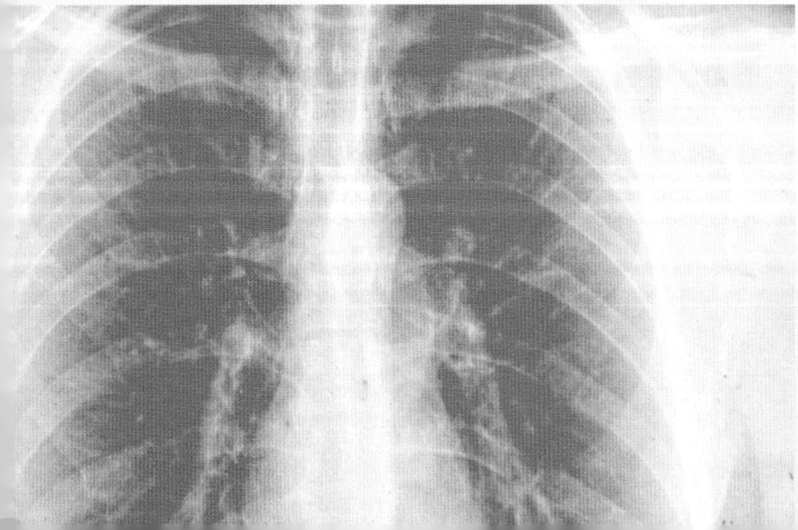
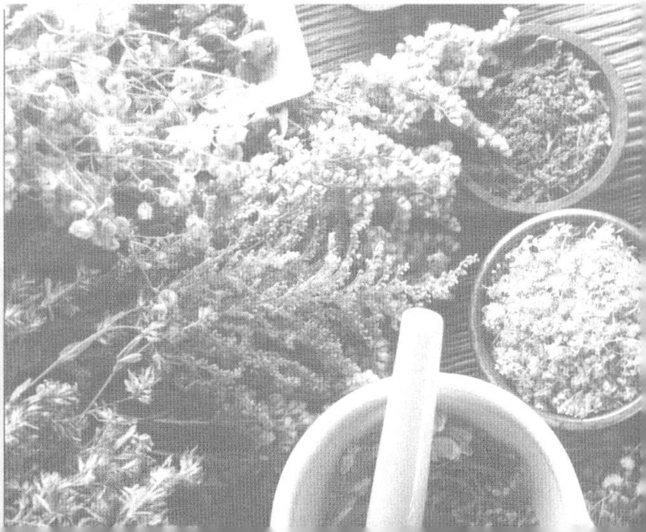

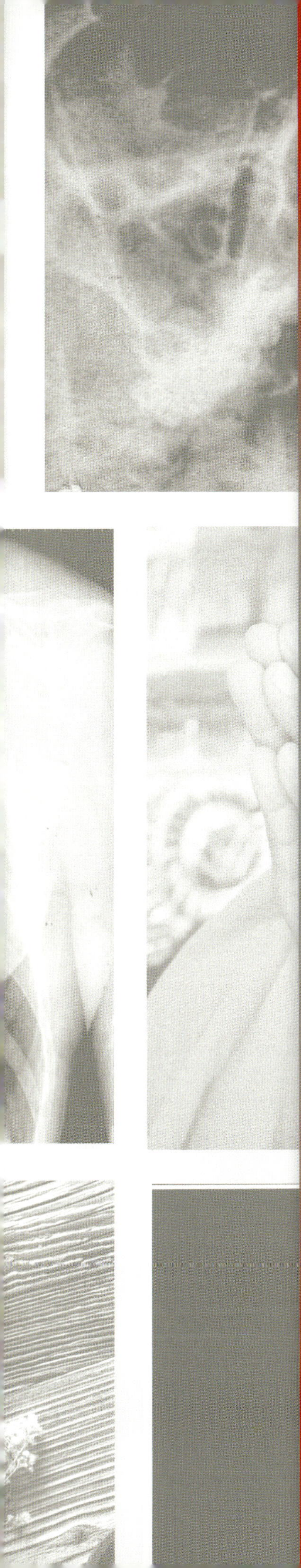

ANCIENT HEALING

Before the advent of modern medicine's imagining, pharmaceutical, and surgical therapies, healers of ancient times practiced a variety of techniques that included herbs, potions, incantations, prayers to the gods and laying-on of hands.

Egyptian manuscripts dating back 4,000 years detail medical centers along the Nile filled with lists of diseases and their treatments. While treatments included medicinal remedies such as castor oil, aloe, mint, cedar, and opium, none was considered effective without magical utterances from a healer. The Egyptians divided the body into 36 parts, each governed by a different god, who needed to directly address the diseased area in order to restore it. Sometimes healers would dress in the form of a particular god to frighten away the disease. Another important element in the healing process required the patient to acknowledge the disease in a spoken statement, which would cause the body, mind, and gods to focus on the cure.

The Greeks appealed to their gods for healing as well through ceremonies, prayers, songs, incantations, and the sacrifice of sacred animals. In their treatments they hoped for divine intervention through a holy object or the laying-on of hands, in which a priest or priestess channeled the power of the divine. Greek healers also instructed their patients to sleep in sacred rooms in order to be healed by their own dreams.

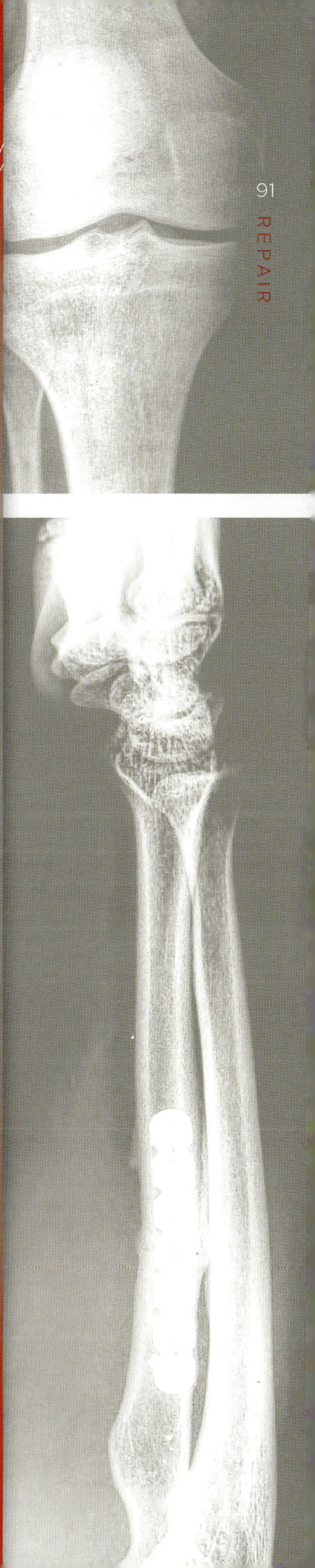

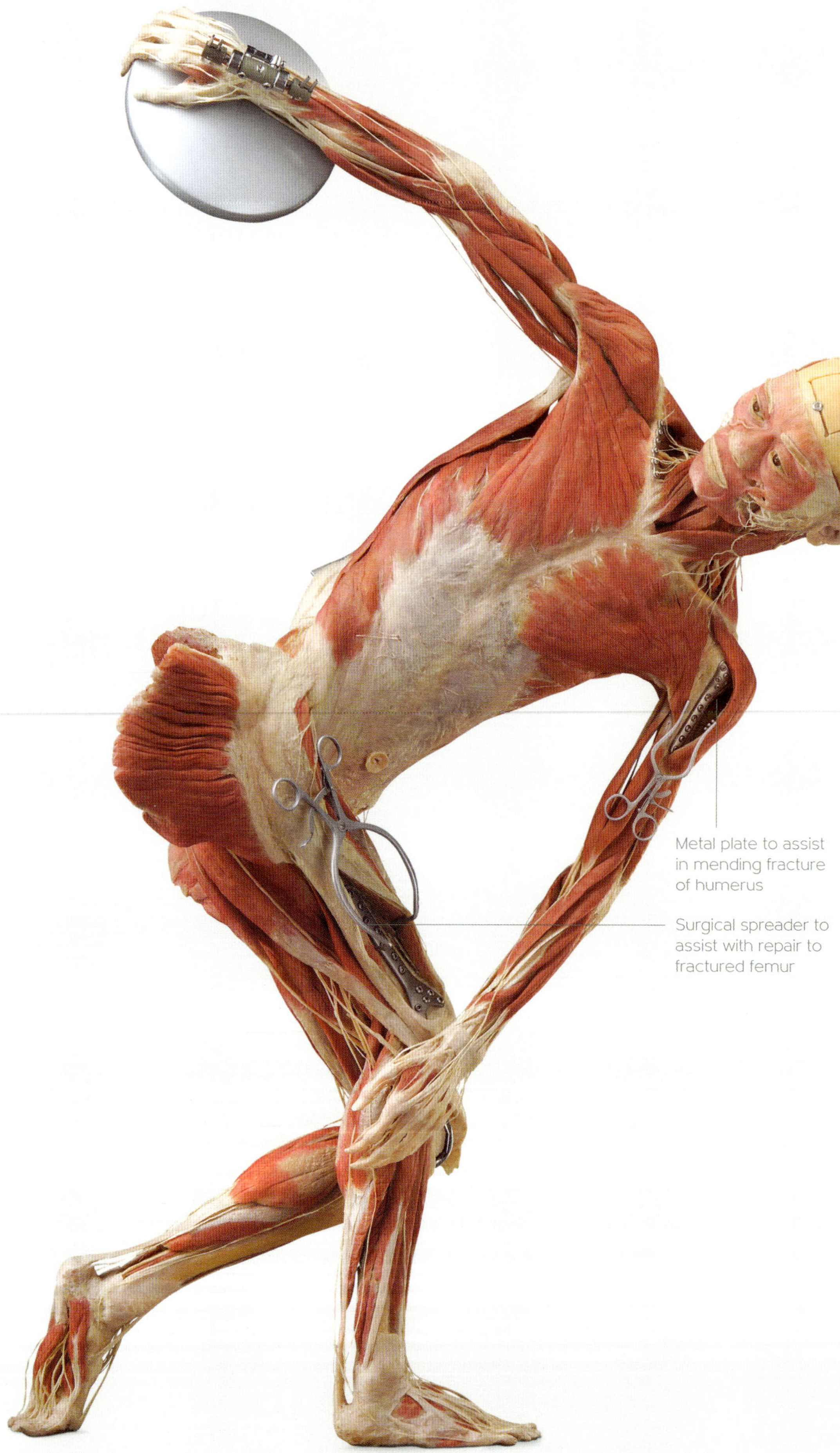
Metal plate to assist in mending fracture of humerus
Surgical spreader to assist with repair to fractured femur

REPAIRING THE BODY

Although a miraculous organism, the body requires maintenance and repair. One example in which repair is required is in the treatment of broken bones. Bone has the remarkable ability to repair itself when immobilized, however, more severe breaks such as a compound fracture or significant fractures of larger bones require treatments such as you see along the femur (thigh bone) of this specimen. Here, the broken bone is set with a steel plate and screws to ensure that the broken ends mend in proper position.

Another common treatment of aging bodies, as well as those with congenital (from birth) defects and injuries, is replacing degenerating or damaged joints. In most cases, joint replacement surgeries decrease pain and restore a tremendous amount of mobility. Advancements are being made in this field each year, leading to longer lasting artificial joints that bond extremely well with bone in the body. This specimen demonstrates examples of hip and knee joint replacements.

The brain is one of the most studied areas of the body and remains a great mystery. However, the physical treatment of the brain for the removal of tumors and stemming cerebral bleeing are quite advanced, as are the methods to open the skull to perform these surgeries. This specimen shows sites where cranial operations occurred.

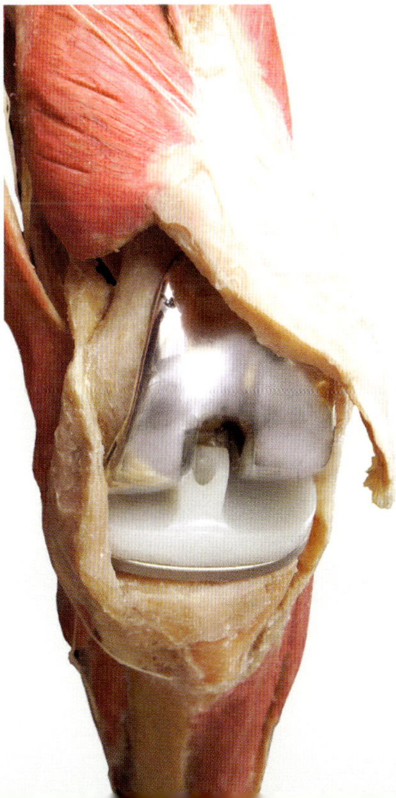

KNEE REPLACEMENT
Damaged or worn out knee joints can now be replaced earlier in life thanks to advancements that allow for longer lasting artificial joints.

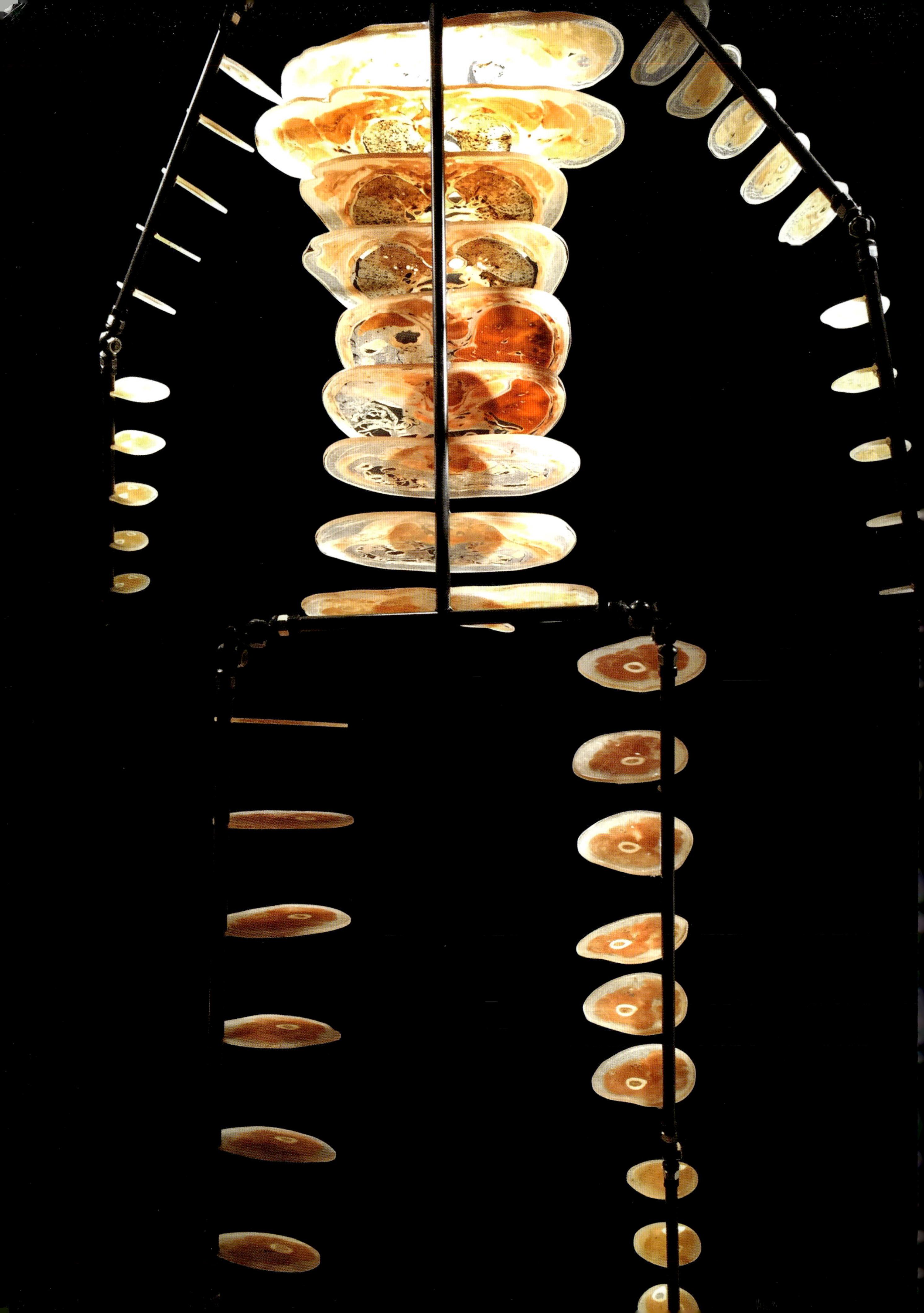

MEDICAL IMAGING

Clinical radiographic techniques such as computerized tomography (CT) and magnetic resonance imaging (MRI) require medical practitioners to understand the anatomical structures and sectional relationships of the body. Researching the spatial relationships of these structures is the key to understanding these scans and diagnosing patients.

This specimen consists of numerous sections of the body regions. The sections displayed here are approximately 3 millimeters (1/8 inch) thick, which is equivalent to some MRI readouts. This specimen was segmented in different ways according to the body's structural features. The torso and the extremities are sectioned horizontally; however, the head, elbows hands, knees and feet are sectioned sagittally. Through these slices, we can easily understand the anatomical structures of different bodily sections.

As this book concludes, this unique presentation offers you an opportunity to see how much of the body you can identify through sectional review. While the entire specimen consists of hundreds of thin sections, only some of these have been selected for display.

> Wherever the art of medicine is loved, there is also a love of humanity.
>
> Hippocrates

WRITTEN IN THE STARS

Since the first human spark emerged in the eyes, we have been studying the heavens. Before history began, the Celtic culture raised massive stones to honor the paths of the sun and moon. 4,000 years ago the priests of Babylon looked toward the skies and performed animal sacrifices to the divine, the mood of their gods represented by the five planets they could track. Around the same time, the Egyptians used 36 groups of stars to mark time. 2,400 years ago, the Chinese Zodiac began to take shape, assigning 12 animals to the trail that Jupiter took through the night sky. The Mayans, Aztecs, Greeks and Romans developed more elaborate systems to explore the pins of light that pierce the night and explain how the planets and stars governed life on Earth.

Our affinity with the stars should not be a surprise because the fundamental atoms that have combined to form us come from the stardust created when the Universe began 14 billion years ago. In fact, some researchers have concluded that more than 90% of our body mass is built of atoms formed by stars. With such a celestial connection, it is no wonder that we believe in the boundless potential of humanity.

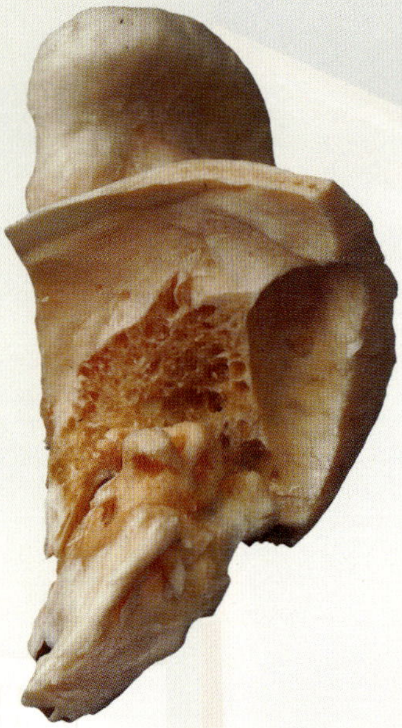

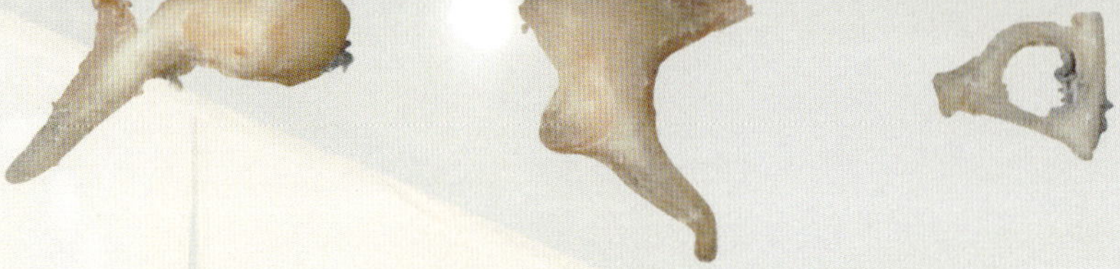

AUDITORY OSSICLES AND INNER EAR
Modern medicine allows us to get a clearer picture of our inner workings than ever before. One area of inquiry has focused on the middle and inner ear. Researchers are looking into the causes and treatment for tinnitus (ringing in the ear) and congenital hearing loss that may involve the auditory ossicles, three small bones in the middle ear cavity whose vibrations are converted into sound.

Research has also focused on the semicircular canals in the middle ear to understand the causes and best treatments for vertigo.

NEED COPY
Impregnation and fetal development occur within the female body; therefore, the principal genital organs such as the uterus, the ovaries, the uterine tubes, and vagina are situated in the pelvic cavity for maximum protection. Only the female pudendum and the accessory genital organ–the mammary glands–lie on the surface of the body.

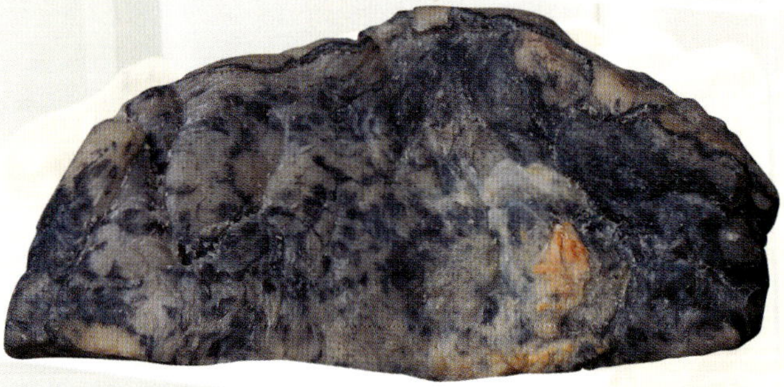

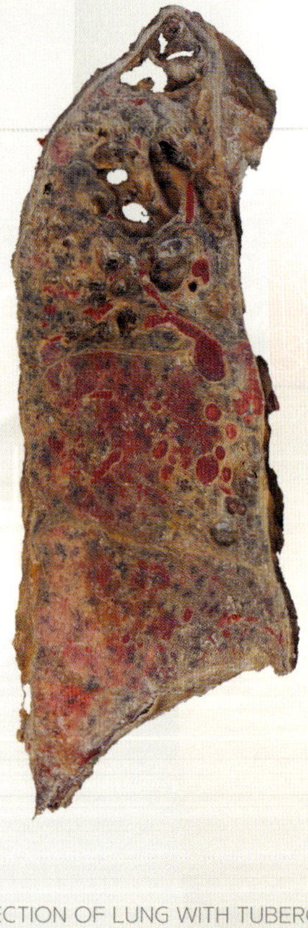

LUNG CANCER
Lung cancer is a leading killer of men and generally occurs in adults aged more than 40. Family history, smoking, and air pollution are risk factors of this disease. Typical symptoms are coughing, blood-streaked sputum, shortness of breath, or chest pain.

SECTION OF LUNG WITH TUBERCULOSIS
Tuberculosis (TB) is an infectious bacterial disease characterized by the growth of nodules in the body, especially the lungs. It spreads through airborne transmission and has a higher occurrence in densely populated areas. Typical symptoms are low fever, night sweat, and a chronic cough. Though highly treatable, TB continues to affect large regions of the developing world.

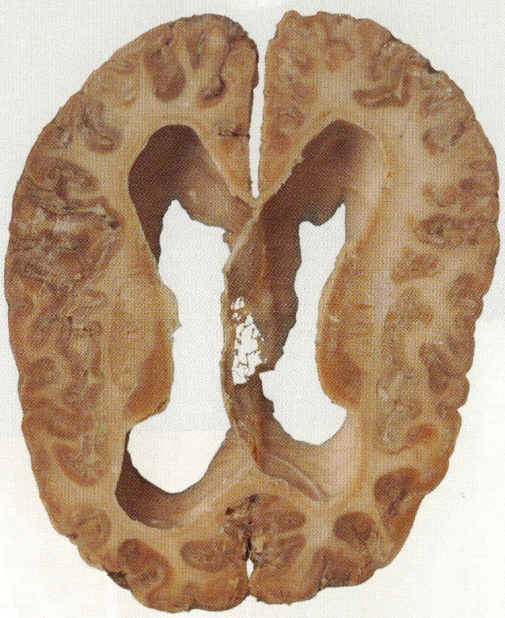

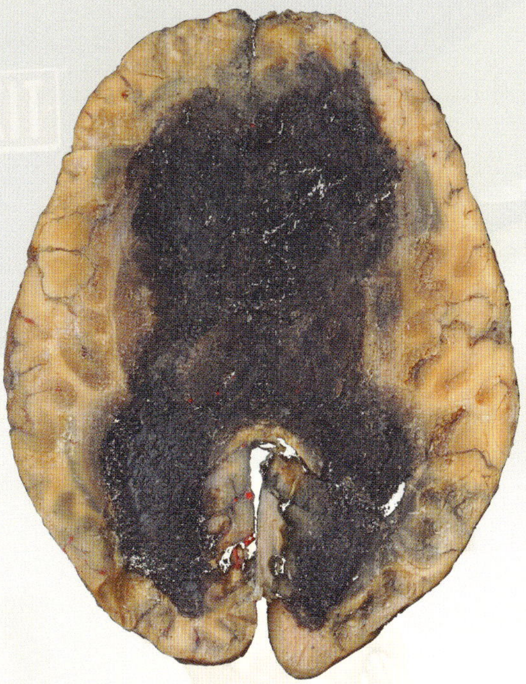

HYDROCEPHALUS
Some diseases can block the flow of cerebrospinal fluid through the ventricles and cause intracranial pressure and expansion of ventricles: a condition known as hydrocephalus.

CEREBRAL BLEEDING
Hypertension, diabetes, and atherosclerosis may induce a rupture of arteries in the brain, which is known as cerebral bleeding or stroke. Blood may put pressure on the brain and result in paralysis, impairment of sensation, and blindness. In serious cases, it will lead to death.

CEREBRAL THROMBOSIS
A thrombosis (blood clot) in the brain can disrupt blood flow to regions of the brain, affecting vision and creating symptoms associated with stroke. If a thrombosis is severe, it can lead to decreased brain function or stroke.

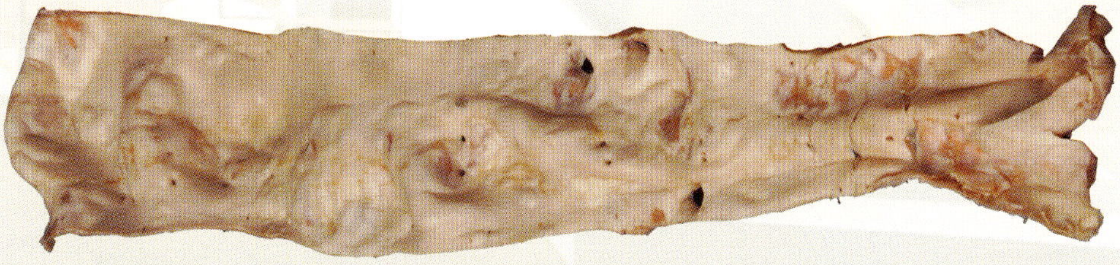

AORTA WITH ATHEROSCLEROSIS
Understanding diet, lifestyle and genetic predisposition helps doctors prevent and treat atherosclerosis more effectively in today's world. High cholesterol, hypertension, smoking, obesity, and mental stress may induce atherosclerosis. When this occurs, the arterial wall loses its normal elasticity and plaques form on it. This blocks the blood flow and can lead to the formation of thrombi (clots). Atherosclerosis may lead to stroke, cerebral thrombosis, and coronary heart disease.

ON REPAIR

The human machine is efficient and productive, capable and rugged, geared for limitless power. And we want to believe our mechanisms are built to last, to carry us ever forward towards a future of our choosing. But no machine is perfect, or runs without interruption–for fuel, inspection, or maintenance.

It is a long road we have set out upon. Most days (we hope) the drive will be smooth; but rough edges and potholes can lead to leaky valves and trembling, creaking, tension, and cracking.

When you need repair, let medicine be your mechanic. From the East and the West thousands of manuals exist that have trained technicians for centuries, and new techniques appear almost daily focused on a better you. Both practical and mysterious, the healing arts are meant for your benefit.

But the greatest doctor, medicine man or shaman lies within. So watch, listen, and hear The Mind and The Gut. They know the treatment best–when to rest, what to crave, when to stop and try again, and when to call upon medicine. Get the maintenance you require–steadily and without hesitation or delay. We want to you to remain among us so you can explore it all–the glory and the pain, the mundane and the magnificent.

It is not lightly that we say Take Care.